DISCRETE STRUCTURES
IN FIVE CHAPTERS

AMIN WITNO

Amin Witno
Department of Basic Sciences
Philadelphia University, Jordan 19392

CreateSpace (July 2010)
ISBN: 1449976611 (pbk.) EAN-13: 9781449976613
Cover image: *The Great Bear Tree* (2008) by Paskal Tanumihardjo.

To our family of five—
Avocado, Eggplant,
Potato, Paprika,
and Peanut

Preface

And further, by these, my son, be admonished: of making many
books there is no end; and much study is a weariness of the flesh.
—King Solomon, quoted in Ecclesiastes 12:12

Discrete Mathematics comprises quite a large number of subdisciplines—
large enough that a semester course can vary greatly from one instructor's
syllabus to another's. Most common elements of Discrete Mathematics are
presented in this small textbook, which has evolved from my Discrete Structures notes throughout sixteen semesters of multi-section classes at Philadelphia University, Jordan. Admittedly, this volume is a product more or less
tailored to fit the expectations in teaching our students here, in terms of style
and choice of contents—in addition to the latter being noticeably inclined
to my taste for number theory.

Guide for the Instructor

The five main chapters are best followed in the given sequence. If desired
otherwise, however, the last three chapters can be arbitrarily permuted without seriously violating the logical chronology of the text.

Exercises are designed to enhance students' understanding of a newly defined concept, stated theorem, or their familiarity with a specific algorithm.
In order to help accomplish their more specific purposes, these problems are
grouped under four different headings:

Exercise x.xx. Regular exercises are essential for the students to retain
material before they proceed to the next, more challenging stage. Although answers, and many complete solutions, are provided, a commit-
ted student should endeavor to solve each problem without resorting
to get help from the back of the book.

Exercise* x.xx. Exercises marked by an asterisk are not necessarily difficult to solve. Rather, being extended workouts which are not absolutely crucial to subsequent results, they may be dealt with at a later time, perhaps assigned at the end of each chapter as a term project. Hence, contrary to what an asterisk symbol may suggest, it is the unmarked exercises that demand immediate attention!

Question. Unnumbered questions mostly have very short answers and are to be asked on the spot during lecture in order to stimulate discussions. Some of these questions may seem trivial while others carry interesting ideas, but each one of them can be a useful test for students' absorption of the subject.

Test x.xx. Multiple-choice questions are rather straightforward and ideally should be discussed at length in class, as students do benefit from learning the arts of taking multiple-choice exams.

It is not too ambitious to cover the entire book in a typical full-semester course. However, in a classroom setting that suits less contents, the further number theoretic topics of Sections 1.4.3–1.4.5, 3.4, 4.3, 4.4.3, may be opted out as independent reading for the enthusiastic students.

Feedback

I will definitely welcome all suggestions, corrections, and comments, whether constructive or otherwise, from users of this text. My gmail account is awitno@gmail.com. Readers are also invited to visit the support web site dedicated to this title, located at

www.witno.com/discrete

to get the latest errata sheet and other product-related updates.

Amman, Summer 2010

Contents

Chapter 1

Topics in Number Theory

We assume familiarity with the number systems. The notion of a number line, which extends from $-\infty$ to $+\infty$, represents the ordering of the *real numbers*. Among these, the counting numbers, 1, 2, 3, ... are better known as the *natural numbers*. There are also the *integers,* which extend the natural numbers by including zero and negative natural numbers. In other words, natural numbers are precisely the positive integers.

The integers come in two kinds, even and odd. The even numbers are

$$0, \ \pm 2, \ \pm 4, \ \pm 6, \ \pm 8, \ \pm 10, \ \pm 12, \ \ldots$$

and the rest of the integers are odd:

$$\pm 1, \ \pm 3, \ \pm 5, \ \pm 7, \ \pm 9, \ \pm 11, \ \pm 13, \ \ldots$$

Observe that the sequence of even numbers can be written in the form

$$\ldots, \ 2 \times -1, \ 2 \times 0, \ 2 \times 1, \ 2 \times 2, \ 2 \times 3, \ \ldots$$

So the two classes of integers may be defined as follows.

Definition. An integer n is *even* if $n = 2m$, where m is again integer. On the other hand, n is *odd* if $n = 2m + 1$ for some integer m.

For instance, 26 is even because $26 = 2 \times 13$, and 13 is integer. But, the number 17 is not even because $17 = 2 \times 8.5$, where 8.5 is not an integer. Note that 17 is odd because $17 = 2 \times 8 + 1$, where 8 is an integer.

Question. With these definitions, can an integer be *both* even and odd? Why or why not?

1

The ratio of two integers, written a/b, with $b \neq 0$, is what we call a *rational number*. Sometimes, a rational number can actually be an integer, e.g., $21/3$ is the integer 7. In general, however, the rational numbers form a bigger set which contains the integers as a subset.

Real numbers which are not rational are called *irrational numbers*. Thus, irrational numbers are real numbers which cannot be expressed as the ratio of two integers. An example of irrational numbers is given by $\sqrt{2}$. We will see in Section 2.3.4 a proof of the fact that $\sqrt{2}$ is indeed irrational.

The integers are the domain of number theory. In particular, number theory is concerned with the properties of the natural numbers. How can we know if a given natural number n is the product of two smaller numbers, which are called *factors* of n? Is there an algorithm to find all common factors of a given pair (m, n)? These are two questions one may ask in number theory.

1.1 Integers in Various Bases

We start by introducing different systems in which we may represent counting numbers. The way we are used to count is based on a ten-digit system, called *decimal*, i.e., using the digits 0 to 9. In computer language, however, it is more convenient to use the *binary* number system, in which we employ only 0 and 1. Computers rely on switches to perceive quantities, and a swith can be *off* or *on*—thus the reason for the binary digits of zeros and ones.

Hence, to enumerate the natural numbers in binary, we begin with

$$1, \ 10, \ 11, \ 100, \ 101, \ 110, \ 111, \ 1000, \ 1001, \ 1010, \ \ldots$$

Note that 111, for instance, corresponds to the number 7 in decimal. We express this relation by writing $111_2 = 7_{10}$. The first question that arises is, given a binary number, how do we know its equivalent in decimal? The key to the algorithm for finding the answer is the following observation.

In decimal, every digit acts as a counter, where from right to left we have the number of ones, then the number of tens (ten ones), the number of hundreds (ten tens), and on. For example,

$$5,467 = 5,000 + 400 + 60 + 7$$
$$= 5 \times 10^3 + 4 \times 10^2 + 6 \times 10^1 + 7 \times 10^0$$

This principle holds in the binary number system as well, except that powers of 10 are replaced by powers of 2. Hence,

$$111_2 = 1 \times 2^2 + 1 \times 2^1 + 1 \times 2^0$$
$$= 4 + 2 + 1$$
$$= 7_{10}$$

Example. Convert the binary number 1100101 to decimal.

Solution. Multiply each digit by the appropriate power of 2, ignoring the zeros since they do not add anything:

$$1100101_2 = 1 \times 2^6 + 1 \times 2^5 + 1 \times 2^2 + 1 \times 2^0$$
$$= 2^6 + 2^5 + 2^2 + 2^0$$
$$= 64 + 32 + 4 + 1$$
$$= 101_{10}$$

Note that without the indices, writing $1100101 = 101$ would have been misleading!

Exercise 1.1. Convert the binary numbers to decimal.
a) 1101111
b) 1110111
c) 100000001
d) 1101111000

Test 1.2. Which binary number represents an odd number?
a) 1100101100
b) 111010010101010
c) 1010010010101011
d) 10101001011111111100

Going in the other direction, how do we convert a decimal number to binary? The reverse algorithm will now involve *divisions* by powers of 2.

Example. Convert the decimal number 101 to binary.

Solution. Ahead of time, we do not know the largest power of 2 which divides into 101. So we will divide 101 by 2 repeatedly, as follows.

$$101 \div 2 = 50 \qquad \text{with remainder } 1$$
$$50 \div 2 = 25 \qquad \text{with remainder } 0$$
$$25 \div 2 = 12 \qquad \text{with remainder } 1$$
$$12 \div 2 = 6 \qquad \text{with remainder } 0$$
$$6 \div 2 = 3 \qquad \text{with remainder } 0$$
$$3 \div 2 = 1 \qquad \text{with remainder } 1$$
$$1 \div 2 = 0 \qquad \text{with remainder } 1$$

Note that the remainders determine the digits of the sought binary number; and we recover the relation $101_{10} = 1100101_2$ by reading these remainders from the last one up.

Exercise 1.3. Convert the decimal numbers to binary.
a) 99
b) 129
c) 999
d) 2730

In principle, the idea of a base-10 or base-2 number system can be generalized to any base-n number system, where n is the number of digits used. Two other common number systems for the computing language are the *hexadecimal* and *octal* systems—using 16 and 8 digits, respectively.

In hexadecimal, we count using the sixteen "digits" 0 to 9 and A to F, in this order. From 1 to 20, for instance, we write,

$$1, \ 2, \ 3, \ 4, \ 5, \ 6, \ 7, \ 8, \ 9, \ A, \ B, \ C, \ D, \ E, \ F, \ 10, \ 11, \ 12, \ 13, \ 14$$

The fact that 14_{16} is equivalent to 20_{10} can be explained by the same principle of "multiples of powers," which has been demonstrated in decimal as well as in binary, i.e., $14_{16} = 1 \times 16^1 + 4 \times 16^0 = 16 + 4 = 20_{10}$.

Example. Convert the hexadecimal number $1A5E$ to decimal.

Solution. This time, multiply each digit by the appropriate power of 16:

$$1A5E_{16} = 1 \times 16^3 + 10 \times 16^2 + 5 \times 16^1 + 14 \times 16^0$$
$$= 4096 + 2560 + 80 + 14$$
$$= 6750_{10}$$

Exercise 1.4. Convert each hexadecimal number given below to decimal.
a) AA
b) CEF
c) $2BAD$
d) 10101

Test 1.5. Which hexadecimal number represents an even number?
a) A625B
b) FF79C3
c) E020ADD
d) 37B951FE

The conversion from decimal to hexadecimal is now through iterative division by 16, analogous to that from decimal to binary.

Example. Convert the decimal number 6750 to hexadecimal.

Solution. We use division with remainder, like in grade school, and find,

$$6750 \div 16 = 421 \qquad \text{with remainder } 14$$
$$421 \div 16 = 26 \qquad \text{with remainder } 5$$
$$26 \div 16 = 1 \qquad \text{with remainder } 10$$
$$1 \div 16 = 0 \qquad \text{with remainder } 1$$

The answer, again, is read from last to first: $6750_{10} = 1A5E_{16}$.

Question. How do we find these remainders in the calculator?

Exercise 1.6. Convert the decimal numbers to hexadecimal.

a) 999
b) 10001
c) 98765
d) 522958

The *octal* number system mentioned earlier employs only the digits 0 to 7. Again, the principles of conversion between two bases remain valid.

Exercise 1.7. Convert the octal numbers to decimal.

a) 777
b) 1234
c) 5702
d) 52543

Exercise 1.8. Convert the decimal numbers to octal.

a) 99
b) 999
c) 10001
d) 98765

The hexadecimal and octal number systems are chosen for the following practical reason. Note the relation $2^4 = 16^1$, which indirectly says that four binary digits, or *bits*, are equivalent to one hexadecimal digit. With the help of Table 1.1, this provides a fast method of conversion between base-16 and base-2.

Example. Convert the hexadecimal number $1A5E$ to binary.

Table 1.1: The numbers 0 to 15 in hexadecimal and in binary.

0	1	2	3	4	5	6	7
0000	0001	0010	0011	0100	0101	0110	0111
8	9	A	B	C	D	E	F
1000	1001	1010	1011	1100	1101	1110	1111

Solution. In fact, we simply replace each hexadecimal digit, 1, A, 5, E, by the corresponding four bits shown in Table 1.1, then we juxtapose these binary digits to form the answer.

$$1A5E_{16} = 0001, 1010, 0101, 1110_2$$

The comas are inserted for better reading, and the answer can well be written without them, i.e., 1101001011110.

Exercise 1.9. Convert the hexadecimal numbers of Exercise 1.4 to binary.

To convert from binary to hexadecimal, simply reverse this action. In the case where the binary digits are not evenly grouped into fours, we simply add extra zeros to the left of the quantity.

Example. Convert the binary number 11011000111100 to hexadecimal.

Solution. There are 14 digits; to group them into fours we need to have two extra zeros on the left. With Table 1.1 again, we get the following answer.

$$0011, 0110, 0011, 1100_2 = 363C_{16}$$

Question. Would it be wrong if we pad zeros to the right?

Exercise 1.10. Convert the binary numbers to hexadecimal.
a) 1101111
b) 11111111111
c) 100000000001
d) 111110000111001

Exercise* 1.11. Elias has carelessly added two extra zeros to the right, instead of to the left of the binary digits and come up with his wrong hexadecimal answer, $ACE8$. What is supposed to be the correct answer?

For base-8, similarly, we have the relation $2^3 = 8^1$. Since this implies that every octal digit corresponds to three bits, there is also a quick way to convert between binary and octal.

Exercise 1.12. Convert the binary numbers to octal, or vice versa.

a) 1101111_2
b) 101110010011_2
c) 264_8
d) 10101_8

Now suppose we wish to convert a hexadecimal number to octal. One way to do this is to convert first to decimal and then to octal—but why not to binary first, and then from binary to octal?

Exercise 1.13. Convert the hexadecimal numbers to octal, or vice versa.

a) $A2C_{16}$
b) $E7DC2_{16}$
c) 5764_8
d) 7777777_8

Test 1.14. We are given a number in the base-4 system, 1231231_4. What is this number in hexadecimal?

a) 6DB1
b) 1B6D
c) 6DB4
d) 1BCD

Exercise* 1.15. The base-26 number system uses the letters of the alphabet, i.e., from A to Z, to represent the digits 0 through 25. How do we represent the decimal number 62534 in base-26?

Exercise* 1.16. Amira is a very wealthy businesswoman who has built a modern village in the suburb of Jakarta. Being superstitious, she refuses to use the digit 4 in numbering the floors of her high-rise office building, the top floor being the 69th. How many floors up is that, if Amira were not afraid to count with 4? Of course, there is no 13th floor either. Can you write a computer program to do this conversion, in either direction?

Appendix: Representing Non-Integer Numbers

We have been concerned with conversion of natural numbers between number systems of different bases. There are ways in which binary digits are used to represent negative integers or even non-integer rational numbers.

In the decimal system, a rational number can be written using a dot (period sign) properly inserted among the digits, e.g., 3.1415. The part to the right of the dot is called the *fractional* part. We observe that the digits of the fractional part represent multiples of negative powers of 10. In this example,

$$3.1415 = 3 + 0.1 + 0.04 + 0.001 + 0.0005$$
$$= 3 \times 10^0 + 1 \times 10^{-1} + 4 \times 10^{-2} + 1 \times 10^{-3} + 5 \times 10^{-4}$$

If we keep this principle for the binary number system, then we may represent certain rational numbers by association with negative powers of 2.

Example. Convert the binary number 0.1011 to decimal.

Solution. We do not need to write down the multiples of zero:

$$0.1011_2 = 2^{-1} + 2^{-3} + 2^{-4}$$
$$= 0.5 + 0.125 + 0.0625$$
$$= 0.6875_{10}$$

Exercise 1.17. Convert the binary numbers to decimal.
a) 0.01
b) 0.10001
c) 0.11111
d) 0.000001

From decimal to binary, converting the fractional part of a rational number would be through repeated multiplication by 2, where in each step we keep record of the integer part.

Example. Convert the decimal number 0.6875 to binary.

Solution. We write the integer parts in the far right column.

$$0.6875 \times 2 = 1 + 0.375 \qquad\qquad 1$$
$$0.375 \times 2 = 0 + 0.75 \qquad\qquad 0$$
$$0.75 \times 2 = 1 + 0.5 \qquad\qquad 1$$
$$0.5 \times 2 = 1 + 0 \qquad\qquad 1$$

This time, the correct answer is obtained by reading the integer parts downward from the top, following the dot: 0.1011_2 .

Note that in the above example we stop the algorithm when we reach 0 in the fractional part. In general, however, the iterations may never terminate with a zero. The situations parallel those in decimal, where a rational number may be represented by an infinite, but always periodic, extension of digits, e.g., $1/3 = 0.333\ldots = 0.\overline{3}$ and $5/11 = 0.\overline{45}$.

Exercise 1.18. Convert the decimal numbers to binary and to hexadecimal.

a) 0.03125
b) 0.765625
c) 5/8
d) 1/3

1.2 Divisibility

We return to the studies of integers. It is clear that adding or multiplying two integers results in another integer. Dividing an integer by another, on the other hand, sometimes yields an integer value but sometimes does not. The relation in which an integer divides another integer (resulting in another integer) is an important concept in the theory of numbers.

First, we need to introduce some functions which have their domain or range in the set of integers. For instance, there are times when we need to extract the integer part of a non-integer number. This particular operation is performed by the floor function.

Definition. The *floor function* $f(x) = \lfloor x \rfloor$ takes any real number x and returns the greatest integer n with condition $n \leq x$. The quantity $\lfloor x \rfloor$ may be called the *floor* of x.

For example, we have $\lfloor 3.1415 \rfloor = 3$ and $\lfloor -100/7 \rfloor = -15$. Note that $\lfloor x \rfloor = x$ if, and only if, x is already an integer. Furthermore, the inequalities $\lfloor x \rfloor \leq x < \lfloor x \rfloor + 1$ hold for any real number x.

A companion to the floor function is the ceiling function, defined as follows.

Definition. The *ceiling function* $f(x) = \lceil x \rceil$ takes any real number x and returns the least integer n with condition $n \geq x$. We may call $\lceil x \rceil$ the *ceiling* of x.

Hence, to illustrate, $\lceil 3.1415 \rceil = 4$ and $\lceil -20/3 \rceil = -6$. Similar to the floor function, we also have $\lceil x \rceil = x$ if and only if x is an integer.

Exercise* 1.19. Order the following quantities, from the smallest to the largest.

$$x, \; \lfloor x \rfloor, \; \lfloor x \rfloor + 1, \; \lfloor x \rfloor - 1, \; \lceil x \rceil, \; \lceil x \rceil + 1, \; \lceil x \rceil - 1$$

1.2.1 The Mod Operation

The floor function is needed to define the next, very useful operation with integers.

Definition. With two integers m and n, where $n > 0$, we define the operation m *mod* n by

$$m \bmod n = m - \left\lfloor \frac{m}{n} \right\rfloor \times n$$

In some programming language, like C++ or Java, the notation $m \bmod n$ is written $m \% n$.

Example. Compute 12345 mod 7.

Solution. According the to definition,

$$12345 \bmod 7 = 12345 - \left\lfloor \frac{12345}{7} \right\rfloor \times 7$$

$$= 12345 - \lfloor 1763.571429\ldots \rfloor \times 7$$

$$= 12345 - (1763 \times 7)$$

$$= 12345 - 12341$$

$$= 4$$

Exercise 1.20. Perform the following mod operations.
a) 678 mod 5
b) 35 mod 217
c) 3393 mod 29
d) 99999 mod 111

Test 1.21. Which one of these four quantities is the largest?
a) 100 mod 7
b) 234 mod 9
c) 11 mod 29
d) 20 mod 11

As you may have suspected by now, the operation $m \bmod n$ actually returns the remainder upon dividing m by n via the division-with-remainder method. In the preceding example, dividing 12345 by 7 will give us the integer output 1763, which we call the *quotient,* and the remainder 4—a fact we may express as an equation,

$$12345 = (1763) \times 7 + (4)$$

The brackets are added merely to emphazise where the quotient and the remainder are, respectively.

The next theorem, whose proof is left as an easy challenge, states some basic properties of the mod operation which are familiar facts concerning the remainder of a division.

Theorem 1.1. Let m and $n > 0$ be fixed integers. Then

1) $0 \leq m \bmod n < n$.

2) $m \bmod n = m$, if $0 \leq m < n$.

3) $m \bmod n = 0$, if m/n is an integer.

4) m/n is an integer, if $m \bmod n = 0$.

The relation $m \bmod n = 0$, appearing in the above theorem, is an important and useful concept in working with integers. This leads us to the next definition.

Definition. The following statements all have one and the same meaning, namely that $m \bmod n = 0$.
a) m is a *multiple* of n
b) m is *divisible* by n
c) n is a *divisor*, or *factor*, of m
d) n *divides* m

In view of Theorem 1.1, this definition also means that m/n is an integer, i.e., there is an integer k such that $m = nk$.

Example. The following examples illustrate the newly defined terms.
a) The fact that $40/8 = 5$, an integer, allows us to say that 8 divides 40 and that 40 is a multiple of 8 or is divisible by 8.
b) The numbers 10, 20, 30, 40, 50, ... are all divisible by 2 and 5.
c) Even numbers are multiples of 2. In contrast, no odd number has a factor of 2.
d) The number 17 has no divisors other than 1 and 17.

Test 1.22. Which number is a multiple of 24?
a) 0
b) 8
c) 16
d) 84

Another important and useful concept involving the mod operation is the relation between integers which have the same remainder upon division by a fixed number $n > 0$.

Definition. If a and b are two integers such that $a \bmod n = b \bmod n$, then we write $a \equiv b \pmod{n}$, and say that a is *congruent* to b mod n. The relation $a \equiv b \pmod{n}$, which is equivalent to $b \equiv a \pmod{n}$, is called a *congruence* mod n.

For example, since 23 mod 7 = 2 and 100 mod 7 = 2, we have $100 \equiv 23$ (mod 7). In this new notation, we can say that $m \equiv 0$ (mod n) precisely when n divides m.

Test 1.23. Which one of the following numbers is congruent to 99 mod 13?
a) 0
b) 69
c) 96
d) 112

1.2.2   An Application in Check Digits

The mod operation is used in many modern applications of identification number assignment, as for a bank account, credit card, airline ticket, product bar code, or a vehicle license plate. In particular, such ID numbers come with a *check digit* (usually the right-most digit) whose purpose is to alert us when an error has occured in typing the number. We illustrate here the use of check digits in assigning the International Standard Book Number (ISBN) for book publications.

An ISBN consists of 10 digits which are separated into four groups by a hyphen between them, e.g., 1-4196-8735-2. These four groups represent the codes for, from left to right, language (0 or 1 means English), publisher, book title, and check digit. In this case, the check digit can also be a capital letter X, and it is determined according to the following algorithm.

Let $a_1, a_2 \ldots, a_{10}$ represent the ten digits of the ISBN, in the order from left to right, and let S be defined by

$$S = (10a_1 + 9a_2 + 8a_3 + 7a_4 + 6a_5 + 5a_6 + 4a_7 + 3a_8 + 2a_9) \bmod 11$$

The check digit will then be given by

$$a_{10} = (11 - S) \bmod 11$$

In addition, due to the range $0 \le a_{10} \le 10$, we agree to replace $a_{10} = 10$ by the letter X.

For example, having determined that the first three codes for the ISBN of a book to be 1-4196-8735-x, we proceed to assigning the check digit x:

$$S = ((10 \times 1) + (9 \times 4) + (8 \times 1) + (7 \times 9) + (6 \times 6)$$
$$+ (5 \times 8) + (4 \times 7) + (3 \times 3) + (2 \times 5)) \bmod 11$$
$$= (10 + 36 + 8 + 63 + 36 + 40 + 28 + 9 + 10) \bmod 11$$
$$= 240 \bmod 11 = 9$$

Thus, $x = (11 - 9) \bmod 11 = 2$, and 1-4196-8735-2 is the complete ISBN.

Exercise 1.24. Determine the check digit for each of the following incomplete ISBN's.

a) 3-314-00783-x
b) 957-747-134-x
c) 962-244-122-x
d) 977-230-154-x

It is not hard to show that the algorithm we have used to produce the check digit a_{10} can be summarized with a single formula,

$$a_{10} = (1a_1 + 2a_2 + 3a_3 + 4a_4 + 5a_5 + 6a_6 + 7a_7 + 8a_8 + 9a_9) \bmod 11$$

It can also be verified that a common typing error like a mistake in just one of the ten digits, or two digits reversely placed, will always be detected by this formula.

Test 1.25. Which one of the following ISBN's is in error?

a) 0-310-91291-1
b) 0-87509-701-4
c) 0-88368-324-X
d) 0-9629049-0-2

Exercise* 1.26. Is it possible, hypothetically, to have two consecutive ISBN's? Think of an example or explain why it is not possible.

As of 1 January 2007, however, all ISBN's have been extended to 13 digits, now called EAN-13, in compliance with the European Article Number for product codes. The conversion is done by prefixing the digit 978 (the code for all books) and readjusting the check digit, according to the following rule.

Let S now be the sum of the first 12 digits, after first multiplying a_2, a_4, a_6, a_8, a_{10}, and a_{12}, each by 3. Then the check digit a_{13} is chosen such that $(S + a_{13}) \bmod 10 = 0$.

Example. Convert the ISBN 1-4196-8735-2 to the corresponding 13-digit EAN.

Solution. The EAN-13 looks like 978-1-4196-8735-x. To determine the check digit, we first calculate S:

$$\begin{aligned}
S &= 9 + (7 \times 3) + 8 + (1 \times 3) + 4 + (1 \times 3) \\
&\quad + 9 + (6 \times 3) + 8 + (7 \times 3) + 3 + (5 \times 3) \\
&= 9 + 21 + 8 + 3 + 4 + 3 + 9 + 18 + 8 + 21 + 3 + 15 \\
&= 122
\end{aligned}$$

Hence we choose the digit $x = 8$ in order to make the sum $122 + 8 = 130$, a multiple of 10. The complete EAN-13 for this book is then 978-1-4196-8735-8 or, as normally written without the hyphens, 9781419687358.

Exercise 1.27. Convert each of the ISBN's in Exercise 1.24 to its corresponding EAN-13.

Question. Do we ever need the letter X in an EAN-13?

Exercise 1.28. A number theory textbook shows on its back cover, ISBN 0-471-62546-9. Elias converted this to EAN-13, ignorantly, by simply adding the prefix: 9780471625469. Kindly correct his answer.

Exercise 1.29. A newly published paperback has 9781449976613 for its EAN-13. What would have been the book's ISBN had it been released before the year 2007?

Exercise* 1.30. Is it possible, hypothetically, to have two consecutive EAN-13's? Think of an example or explain why it is not possible.

1.2.3   GCD and LCM

With two integers, it is useful sometimes to find a divisor common to both. For example, 5 is a divisor of both 10 and 25. The next theorem says something about a property of a common divisor.

Theorem 1.2. Suppose d is a divisor of both m and n. Then d divides $am + bn$ for any integers a and b.

Proof. Since m/d and n/d are both integers, the number

$$\frac{am + bn}{d} = a \times \frac{m}{d} + b \times \frac{n}{d}$$

is also an integer, if a and b are.   ▽

Now, given an integer $m \neq 0$, there exist only a finite number of divisors. This is so because if m/n is an integer then $|n| \leq |m|$. The next function will take two integers m and n and returns the greatest of all divisors common to both.

Definition. Let m and n be two integers, not both zero. The *greatest common divisor* of m and n is the largest integer d which divides both m and n. We shall denote this quantity by writing $d = \gcd(m, n)$.

For example, $\gcd(18, 30) = 6$ because 6 divides both 18 and 30, and 6 is the largest number with such a property.

Exercise 1.31. Evaluate $\gcd(m, n)$ given below.

a) $\gcd(125, 200)$
b) $\gcd(12345, 0)$
c) $\gcd(-12, 145)$
d) $\gcd(2, 10000)$

The following theorem will be essential in evaluating $\gcd(m, n)$ for arbitrary values of m and n, even if they are very large.

Theorem 1.3. We have $\gcd(m, n) = \gcd(n, m \bmod n)$.

Proof. Any common divisor of m and n also divides $m \bmod n = m - \lfloor m/n \rfloor n$ by Theorem 1.2. Conversely, any common divisor of n and $m \bmod n$ also divides $m = m \bmod n + \lfloor m/n \rfloor n$ by the same theorem. Hence, both pairs (m, n) and $(n, m \bmod n)$ share the same set of all divisors common to them and, in particular, equal common divisor of greatest value. ▽

Applying Theorem 1.3 twice gives us $\gcd(m, n) = \gcd(m \bmod n, n \bmod (m \bmod n))$. By iteration, the pair decreases in size quite rapidly. This iterative procedure is called the *Euclidean algorithm,* a very efficient method for computing gcd.

Example. Evaluate $\gcd(12345, 6789)$ by the Euclidean algorithm.

Solution. Repeated application of Theorem 1.3 allows us to write

$$\begin{aligned}
\gcd(12345, 6789) &= \gcd(6789, 5556) &&\text{since } 12345 \bmod 6789 = 5556 \\
&= \gcd(5556, 1233) &&\text{since } 6789 \bmod 5556 = 1233 \\
&= \gcd(1233, 624) &&\text{since } 5556 \bmod 1233 = 624 \\
&= \cdots
\end{aligned}$$

Or we may opt to write only the sequence of remainders:

$$12345, \ 6789, \ 5556, \ 1233, \ 624, \ 609, \ 15, \ 9, \ 6, \ 3, \ 0$$

The last pair tells us that $\gcd(12345, 6789) = \gcd(3, 0) = 3$.

Question. Does the Euclidean algorithm always terminate with a zero remainder?

Exercise 1.32. Use the Euclidean algorithm to evaluate $\gcd(m, n)$.

a) $\gcd(12345, 67890)$
b) $\gcd(12345, 54321)$
c) $\gcd(88888, 555)$
d) $\gcd(234, 60970)$

We conclude this section with one more integer function which complements the gcd function, i.e., the least common multiple.

Definition. With positive integers m and n, we define their *least common multiple* to be the least positive integer which is divisible by both m and n, denoted by $\text{lcm}(m, n)$.

We have $\text{lcm}(12, 15) = 60$, for instance, since 60 is a common multiple of 12 and 15, and it is the smallest of such.

We do not have a particular algorithm to evaluate $\text{lcm}(m, n)$, but the following equality reveals a nice relation between $\gcd(m, n)$ and $\text{lcm}(m, n)$ which can well be used to evaluate one given the other.

Theorem 1.4. For positive integers m and n, we have

$$\gcd(m, n) \times \text{lcm}(m, n) = m \times n$$

We postpone the proof of this claim until later when we reestablish this result following Theorem 1.9 in this chapter.

For example, to evaluate $\text{lcm}(12, 15)$ we may first note that $\gcd(12, 15) = 3$, from which we conclude that $\text{lcm}(12, 15) = 12 \times 15/3 = 60$.

Exercise 1.33. Evaluate $\text{lcm}(m, n)$ by first evaluating $\gcd(m, n)$.

a) $\text{lcm}(275, 115)$
b) $\text{lcm}(144, 456)$
c) $\text{lcm}(999, 123)$
d) $\text{lcm}(725, 1000)$

1.3   Solving Linear Equations

Given integers m, n, and c, we are interested in finding solutions to the linear equation in two variables, x and y, of the form

$$mx + ny = c \tag{1.1}$$

By solutions we mean integer solutions. It turns out that the main ingredient in solving equations of this kind is in fact the Euclidean algorithm.

Theorem 1.2 reminds us that if d is a common divisor of m and n, then d divides $mx + ny$ for any integer values of x and y. Therefore, the first condition for Equation (1.1) to have a solution is that c must be divisible by d and, in particular, by $\gcd(m, n)$.

Theorem 1.5. If the linear equation $mx + ny = c$ has a solution for x and y which are both integers, then $\gcd(m, n)$ must divide c.

Conversely, when c is a multiple of $\gcd(m, n)$, we claim that integer solutions x and y for Equation (1.1) always exist. How do we find at least one such solution pair? First, we claim that integers a and b exist such that

$$ma + nb = \gcd(m, n) \tag{1.2}$$

Then if $c/\gcd(m, n)$ is an integer, we multiply through Equation 1.2 by this integer to obtain

$$m\left(\frac{ac}{\gcd(m, n)}\right) + n\left(\frac{bc}{\gcd(m, n)}\right) = c$$

thereby producing a solution (x, y) for Equation (1.1).

And how do we find an integer pair (a, b) for Equation (1.2)? We need an extension of the Euclidean algorithm, thus called the *extended Euclidean algorithm*, which we illustrate in the next example.

Example. Find integers a and b such that $123a + 45b = \gcd(123, 45)$.

Solution. We start by writing rows of three integers, labeled (d_i, a_i, b_i) for each row $i \geq 1$, beginning with

	d_i	a_i	b_i
1	123	1	0
2	45	0	1

To determine the third row, subtract $\lfloor 123/45 \rfloor = 2$ times the entire second row from the first. In particular, we will have $d_3 = 123 - \lfloor 123/45 \rfloor 45 = 123 \bmod 45 = 33$. Similarly, for the fourth row, we substract $\lfloor 45/33 \rfloor = 1$ times the entire third row from the second, so that $d_4 = 45 \bmod 33 = 12$.

In this way, down the first column we have the sequence of remainders which we would have upon computing $\gcd(123, 45)$ using the Euclidean algorithm, i.e.,

$$123, 45, 33, 12, 9, 3, 0$$

The completed table with the seven rows is thus obtained:

	d_i	a_i	b_i
	123	1	0
(-2)	45	0	1
(-1)	33	1	-2
(-2)	12	-1	3
(-1)	9	3	-8
(-3)	3	-4	11
	0	15	-41

In such table, we claim that each row obeys the relation

$$d_i = 123a_i + 45b_i \qquad (1.3)$$

In particular, the row before the last gives us $\gcd(123, 45) = 3 = 123(-4) + 45(11)$. Thus, we have found our solution of $a = -4$ and $b = 11$.

Question. Can you *prove* why the relation (1.3) holds in each row?

Exercise 1.34. For each given pair (m, n), find integers a and b such that $ma + nb = \gcd(m, n)$.
a) $(345, 215)$
b) $(826, 112)$
c) $(2890, 843)$
d) $(529, 6739)$

Example. Find integers x and y such that $123x + 45y = 66$.

Solution. In the last example, we have found that $123(-4) + 45(11) = 3$. Simply multiply by $66/3 = 22$, and we have a particular solution $x = -88$ and $y = 242$.

Test 1.35. Which equation has integer solutions?
a) $12x + 27y = 35$
b) $12x + 27y = 15$
c) $12x + 20y = 35$
d) $12x + 20y = 15$

Exercise 1.36. For each pair (m, n) given in Exercise 1.34, find integers x and y such that $mx + ny = c$.
a) $c = 95$
b) $c = 98$
c) $c = 11$
d) $c = 99$

In remark, a solution pair for (1.1) in general is not unique. In the preceding table, for instance, if we multiply the third row by 2, then $123(2) + 45(-4) = 66$, providing another solution pair to $123x + 45y = 66$.

Under the condition that $\gcd(m, n)$ divides c, it is now established that Equation (1.1) has at least one solution, or a *particular solution*, denoted by (x_0, y_0). The next theorem describes how to find *all* the solutions.

Theorem 1.6. The equation $mx + ny = c$ has a solution if and only if $\gcd(m, n)$ divides c, in which case all its solutions are given in the form

$$x = x_0 - \frac{nk}{\gcd(m, n)} \quad \text{and} \quad y = y_0 + \frac{mk}{\gcd(m, n)}$$

for any particular solution (x_0, y_0) and for any integer k.

Proof. If we were working over the real numbers, the solutions to $mx+ny = c$ would be represented by a straight line passing through the point (x_0, y_0) and with a slope equals $-m/n$. An arbitrary point on this line is therefore given by (x, y), where

$$x = x_0 - t \quad \text{and} \quad y = y_0 + tm/n$$

for any real number t. We want points on this line which have integer coordinates, so we require that both t and tm/n be integers. We leave as an exercise to verify that this desired condition is achieved precisely when t is a multiple of $n/\gcd(m, n)$, in order to complete the proof.   ▽

Example. Find *all* integers x and y such that $123x + 45y = 66$.

Solution. Since we have found a particular solution $(-88, 242)$, and since $\gcd(123, 45) = 3$, the general solutions are now given by

$$x = -88 - 15k \quad \text{and} \quad y = 242 + 41k$$

for any integer k. For example, with $k = -6$ we have the particular solution $x = 2$ and $y = -4$, of which we have remarked earlier.

Exercise 1.37. Complete Exercise 1.36 by finding the general solutions.

Exercise* 1.38. Elias placed a take-out order from Tea Kitchen Chinese restaurant, where a bowl of seafood fried rice costs 3 dinars, a plate of General Tso's chicken is 5.5 dinars, and individually wrapped spring rolls sell for 20 piasters (0.2 dinar) a piece. Elias spent exactly 100 dinars, and he remembered there were exactly 100 items in the bag. Can you break down the receipt for him?

1.4 Prime Numbers and Factorization

The term *factorization* refers to the process of expressing a positive integer as the product of two smaller numbers. For instance, we *factor* the number 91 when we write $91 = 7 \times 13$. In this sense, factorization is the reverse action of multiplication. A prime number can be thought of as an integer which cannot be factored. More precisely,

Definition. An integer $p \geq 2$ is called *prime* or a *prime number,* if it has no divisor strictly between 1 and p. The list of prime numbers begins with

$$2, \ 3, \ 5, \ 7, \ 11, \ 13, \ 17, \ 19, \ 23, \ 29, \ 31, \ 37, \ 41, \ 43, \ \ldots$$

An integer $n \geq 2$ which is not prime is called a *composite.*

1.4.1   Unique Factorization into Primes

Prime numbers are the building blocks of the integers, in the sense that every integer can be written as a product of primes, and that in an essentially unique way. We now observe some properties of primes which will lead to the establishment of this claim.

Theorem 1.7. Let p be a prime number. If p divides a product of integers, then p must divide one of them.

Proof. Assume that p divides mn. If p does not divide m, we will show that p divides n. Look at $\gcd(m, p)$. Being a divisor of p, this quantity is either 1 or p. So, if p does not divide m, then $\gcd(m, p) = 1$. Using the extended Euclidean algorithm, we can find integers a and b such that $ma + pb = 1$. Multiply this equality by n/p to get

$$\left(\frac{mn}{p} \right) a + nb = \frac{n}{p}$$

The quantity on the left-hand side is an integer, since p divides mn; hence so is n/p an integer, i.e., p divides n.

This argument can be repeated to prove the theorem for the case where the product involves three integers or more.   ▽

By definition, a composite n can be factored as $n = a \times b$, where $1 < a, b < n$. If a or b is again composite, we can factor it again, and again, and with each step the factors decrease in size. After a finite number of steps, the final stage in this process will be something like

$$n = p_1 \times p_2 \times p_3 \times \cdots \times p_k$$

where each p_i is a prime number. But is it possible, for the same n, another person comes down to a *different* factorization, other than mere reordering of the primes? Well, suppose there are two such results:

$$p_1 \times p_2 \times p_3 \times \cdots \times p_k = q_1 \times q_2 \times q_3 \times \cdots \times q_h \qquad (1.4)$$

We may cancel off common primes from each left and right, and if the p's and the q's are really different, then we end up with an equality like (1.4) in which none of the p's equals any of the q's.

However, by Theorem 1.7, p_1 must divide one of the q's. This cannot happen as distinct primes do not divide each other. And that can only mean that the factorization of n into primes involves a unique collection of prime factors. We have proved the *fundamental theorem of arithmetic*.

Theorem 1.8 (The Fundamental Theorem of Arithmetic). Every integer $n \geq 2$ can be factored into prime numbers in a unique way, apart from reordering of the prime factors.

For example, in factoring the number 936 into primes, one may obtain $936 = 3 \times 2 \times 3 \times 13 \times 2 \times 2$, while another $936 = 13 \times 2 \times 2 \times 3 \times 2 \times 3$. But, it would be impossible to find a prime factor outside the collection $\{2, 2, 2, 3, 3, 13\}$. We normally write the final factorization, where all the factors are primes, using the exponential notation in order to clearly display the repeated primes, e.g.,

$$936 = 2^3 \times 3^2 \times 13 \tag{1.5}$$

Exercise* 1.39. Amira claims that she has found a counter-example to the fundamental theorem of arithmetic by showing a different factorization: $936 = 2 \times 3 \times 167$, which she insists is correct—only that it is not written in decimal. Which integer base does Amira have in mind? Does her finding really contradict Theorem 1.8?

Exercise 1.40. Factor these numbers into primes.
a) 888
b) 36000
c) 63756
d) 111111

We have sensed here that factoring in general is harder than multiplying. The most basic, and slowest, factoring algorithm is the *trial division*, where we repeatedly divide n by the primes 2, 3, 5, 7, ... in an attempt to discover a prime factor. Note that only primes up to $\sqrt{n}$ need to be considered, and if no such factor is found then we may conclude that n is itself prime.

Question. Why don't we need to consider prime factors larger than $\sqrt{n}$?

Example. Determine whether the number 577 is prime or composite, using trial division.

Solution. We have $\sqrt{577} \approx 24.02$. The only prime numbers up to 24 are 2, 3, 5, 7, 11, 13, 17, 19, and 23. After a little bit of checking, we see that none of these primes divides 577. Hence, 577 is itself a prime number.

Exercise 1.41. Determine prime or composite by trial division. If composite, factor the number into primes.
a) 239
b) 841
c) 911
d) 1147

1.4.2   GCD and LCM via Factorization

It is sometimes convenient to express the factorization of n into primes using the product notation,

$$n = \prod_{i \geq 1} p_i^{e_i} = p_1^{e_1} \times p_2^{e_2} \times p_3^{e_3} \times \cdots$$

where the product ranges over all prime numbers, $p_1 = 2$, $p_2 = 3$, $p_3 = 5$, $\ldots$ The exponents e_i will be zero except for finitely many of them. For example with $n = 936$, Equation (1.5) shows that $e_1 = 3$, $e_2 = 2$, $e_6 = 1$, and $e_i = 0$ for all the rest.

Now let $d = \prod p_i^{d_i}$ be the prime factorization of another integer d. As a consequence of Theorem 1.7 and the fundamental theorem of arithmetic, if d divides $n = \prod p_i^{e_i}$ then it is necessary that $d_i \leq e_i$. This fact leads to a more explicit way of evaluating the functions $\gcd(m, n)$ and $\text{lcm}(m, n)$, provided that the factorizations of m and n have been established.

Theorem 1.9. Let $m = \prod p_i^{f_i}$ and $n = \prod p_i^{e_i}$, and let $\min(e_i, f_i)$ and $\max(e_i, f_i)$ denote the lesser and the greater, respectively, of e_i and f_i. Then,

$$\gcd(m, n) = \prod p_i^{\min(e_i, f_i)} \quad \text{and} \quad \text{lcm}(m, n) = \prod p_i^{\max(e_i, f_i)}$$

Proof. Suppose d divides n, where $d = \prod p_i^{d_i}$. We have argued that a necessary condition for d is that $d_i \leq e_i$. Now if d also divides m then as well $d_i \leq f_i$. Hence the greatest common divisor of m and n is such an integer d for which $d_i = e_i$ or $d_i = f_i$, whichever is smaller. This gives $d_i = \min(e_i, f_i)$. The proof for $\text{lcm}(m, n)$ is very similar.   ▽

For example, having factored $m = 2^5 \times 3 \times 7^2 \times 13^8 \times 37 \times 101$ and $n = 2^{11} \times 3^2 \times 5^9 \times 11 \times 13^4 \times 23 \times 37$, we readily conclude that

$$\gcd(m, n) = 2^5 \times 3 \times 13^4 \times 37$$
$$\text{lcm}(m, n) = 2^{11} \times 3^2 \times 5^9 \times 7^2 \times 11 \times 13^8 \times 23 \times 37 \times 101$$

Thus Theorem 1.9 provides a second method for evaluating $\gcd(m, n)$ without the use of the Euclidean algorithm. Even so, factorization in general is extremely time-consuming—while in contrast, the Euclidean algorithm is particularly a very efficient algorithm.

Exercise 1.42. Redo Exercise 1.32, this time evaluate $\gcd(m, n)$ by factoring m and n.

Note that with Theorem 1.9, we are now able to derive the relation

$$\gcd(m,n) \times \text{lcm}(m,n) = m \times n \tag{1.6}$$

thereby proving Theorem 1.4. Details are asked in the next exercise.

Exercise 1.43. Redo Exercise 1.33, this time evaluate both $\gcd(m,n)$ and $\text{lcm}(m,n)$ by way of factoring m and n. In each case, verify that (1.6) holds and then try to write a proper proof of Theorem 1.4.

As a further consequence of the fundamental theorem of arithmetic, it is not difficult to show that the list of prime numbers never ends. This claim is stated in the next theorem, whose proof was first given by Euclid some 2500 years ago. Our proof here is a slightly modified version of his.

Theorem 1.10. There are infinitely many prime numbers.

Proof. Form the following sequence of integers.

$$a_1 = 2$$
$$a_2 = a_1 + 1 = 3$$
$$a_3 = a_1 a_2 + 1 = 7$$
$$a_4 = a_1 a_2 a_3 + 1 = 43$$
$$\vdots$$
$$a_k = a_1 a_2 a_3 \cdots a_{k-1} + 1$$

We claim that every pair (a_m, a_n) of two numbers taken from this sequence has $\gcd(a_m, a_n) = 1$. To see why this is true, assuming $m > n$, we can write

$$a_m = a_1 a_2 a_3 \cdots a_n \cdots a_{m-1} + 1$$

which shows that $a_m \bmod a_n = 1$. By Theorem 1.3, we have $\gcd(a_m, a_n) = \gcd(a_n, 1) = 1$. This says that each successive term in the sequence a_k yields a completely new set of prime factors, proving their infinitude. $\triangledown$

Question. Does this theorem imply that there are only finitely many composites?

1.4.3 Power Mod Computations

We are to observe that in computing $ab \bmod n$, we may first reduce a and b by replacing them with their respective remainders mod n. The congruence notation, which was introduced in Section 1.2.1, provides a convenient way to state this theorem.

Theorem 1.11. Let $n > 0$ be a fixed integer. For any integers a and b,

$$(a \bmod n)(b \bmod n) \equiv ab \pmod{n}$$

Proof. By definition, we have

$$(a \bmod n)(b \bmod n) = \left(a - \lfloor \frac{a}{n} \rfloor n \right) \left(b - \lfloor \frac{b}{n} \rfloor n \right)$$

$$= ab + n \left(\lfloor \frac{a}{n} \rfloor \lfloor \frac{b}{n} \rfloor n - a \lfloor \frac{b}{n} \rfloor - b \lfloor \frac{a}{n} \rfloor \right)$$

Moreover, since $ab = ab \bmod n + \lfloor ab/n \rfloor n$, we are then allowed to write

$$(a \bmod n)(b \bmod n) = ab \bmod n + nk$$

for some integer k. It follows that $(a \bmod n)(b \bmod n) \bmod n = ab \bmod n$, proving the congruence.   ▽

Now some applications, like in cryptography, involve the task of computing an expression of the form $a^k \bmod n$ with a very large exponent k, e.g., $2^{1000} \bmod 7$. Note that in this example, while the power 2^{1000} is quite large, its remainder mod 7 will not exceed 6!

With Theorem 1.11, we will evaluate $a^k \bmod n$ by iteratively multiplying a to itself, k times, while in each step reducing the product mod n, in order to keep the calculations manageable. Being more clever, the *successive squaring algorithm*, described next, achieves this goal in much less time.

Example. Compute $3^{234} \bmod 25$ by the successive squaring algorithm.

Solution. We will form a sequence of successive squares with initial term 3, in which each term is reduced mod 25. In displaying the result below, we omit writing "mod 25" for better readability.

$$3^2 = 9$$
$$3^4 = 9^2 = 6$$
$$3^8 = 6^2 = 11$$
$$3^{16} = 11^2 = 21$$
$$3^{32} = 21^2 = 16$$
$$3^{64} = 16^2 = 6$$
$$3^{128} = 6^2 = 11$$

The next square, 3^{256}, is bigger than 3^{234}, so we stop here. Next, we express the exponent 234 in binary, which is really the sum of powers of 2, i.e.,

$$234 = 11101010_2 = 128 + 64 + 32 + 8 + 2$$

Finally, we rely on Theorem 1.11 to conclude that

$$3^{234} \bmod 25 = (3^{128} \times 3^{64} \times 3^{32} \times 3^8 \times 3^2) \bmod 25$$
$$= (11 \times 6 \times 16 \times 11 \times 9) \bmod 25 = 19$$

Exercise 1.44. Use the successive squaring algorithm for each power mod.
a) $2^{22} \bmod 10$
b) $5^{99} \bmod 36$
c) $23^{333} \bmod 100$
d) $2^{2249} \bmod 23$

Test 1.45. What is the *unit digit*, i.e., right-most digit, of the number 7^{99}?
a) 1
b) 3
c) 7
d) 9

From the theoretical point of view, power mod operation touches on an elegant theorem of Fermat, which plays an important role in the RSA cryptography of the next section. However, the theorem will not be proved until later in the text—twice, in fact, restated as Theorems 3.30 and 4.15.

Theorem 1.12 (Fermat's Little Theorem). Suppose that a is an integer not divisible by the prime p. Then $a^{p-1} \bmod p = 1$.

For example, knowing that 5647 is prime, Fermat's little theorem assures us that $89^{5646} \bmod 5647 = 1$.

Exercise 1.46. Compute the following powers mod 23, a prime, mentally—with the help of Fermat's little theorem.
a) $100^{22} \bmod 23$
b) $5^{24} \bmod 23$
c) $3^{47} \bmod 23$
d) $2^{2249} \bmod 23$

Exercise* 1.47. If p is a prime number, prove that $a^p \equiv a \pmod{p}$ for every integer a.

Exercise 1.48. Prove that 779 is composite without factoring it, but by showing that $2^{778} \bmod 779 \neq 1$, thereby failing the statement of Fermat's little theorem.

Exercise* 1.49. Is it possible to have $2^{p-1} \bmod p = 1$, but p is composite? Find an example or explain why it is not possible.

1.4.4   An Application in Cryptography

The technology of data transfer has become an inseparable part of the modern life, be it over the Internet, email, or mobile telephone. At times it becomes necessary to send sensitive data, such as a credit card number, over a secure line.

Cryptography is a field of study wherein we analyze different algorithms by which we convert such a sensitive numeric into a secret number which can be read only by the intended recipient who possesses the secret key to it. (A non-numerical message can be treated numerically, usually by assigning a value to each character such as that based on the ASCII table.) One application we wish to present here is the RSA algorithm, named after its three inventors, Rivest, Shamir, and Adleman in 1976.

Let's say Amira represents an online company which involves receiving important data from its users. She secretly selects two distinct, very large prime numbers p and q (of at least 100 digits each) and another positive integer e such that

$$\gcd((p-1)(q-1), e) = 1$$

Of course, Amira uses the Euclidean algorithm to check this gcd condition. In fact, she employs the extended Euclidean algorithm, which gives her two more integers, $a < 0$ and $b > 0$, such that

$$(p-1)(q-1)a + eb = 1$$

Question. What if the algorithm does not yield $a < 0$ and $b > 0$?

Amira then computes $n = p \times q$ and goes on to post on her web site the two values of n and e, with the following instruction: Everyone who wishes to send her an integer m (the sensitive message) must first convert m into a secret number s, based on the formula

$$s = m^e \bmod n$$

This can be performed efficiently using the successive squaring algorithm. And when Amira receives this value of s, she uses her secret key b to recover the intended message m, also using the successive squaring algorithm, i.e.,

$$s^b \bmod n = m \tag{1.7}$$

Why is this true? First, by Fermat's little theorem, we have

$$m^{(p-1)(q-1)} \bmod p = (m^{p-1})^{q-1} \bmod p = 1^{q-1} \bmod p = 1$$
$$m^{(p-1)(q-1)} \bmod q = (m^{q-1})^{p-1} \bmod q = 1^{p-1} \bmod q = 1$$

These two equations imply that $m^{(p-1)(q-1)} - 1$ is a common multiple of p and q. Being distinct, both p and q must appear in the prime factorization of $m^{(p-1)(q-1)} - 1$. Hence, $m^{(p-1)(q-1)} - 1$ is actually a multiple of pq, and

$$m^{(p-1)(q-1)} \bmod pq = 1$$

Remembering that $p \times q = n$, we observe that

$$m^{eb} = m^{1-(p-1)(q-1)a} = m \times (m^{(p-1)(q-1)})^{-a}$$

and therefore, proving (1.7),

$$s^b \bmod n = m^{eb} \bmod n = m(1)^{-a} \bmod n = m$$

assuming that $m < n$. With the size of n being very large, this is probably the case, but if $m > n$ then m needs to be cut up into two or more blocks of smaller integers and sent one at a time.

Question. Where has Theorem 1.11 been used again in this algorithm?

However, just how secure is this RSA algorithm? Recall that only n and e are known to the public. In the worst case, an enemy can also steal s when it is transmitted across the Internet. Knowing n, e, and s, can the enemy recover the secret key b and/or the intended message m?

The only known feasible way to retrieve b is to first find the factors p and q; and that is exactly the strength of RSA: While multiplying takes a quadratic time, with respect to the number of digits in p and q, factoring takes an exponential time. To illustrate, with the size of n around 200 digits, if multiplying p and q took only one second, then factoring n would take 10^{18} years!

Example. Let us suppose, for a small example, that $p = 29$ and $q = 101$. Hence, $n = 29 \times 101 = 2929$ and $(p-1)(q-1) = 2800$. Amira selects $e = 13$ and runs the extended Euclidean algorithm, arriving at the result

$$2800(-5) + 13(1077) = 1$$

Her secret key is $b = 1077$, whereas the values of $n = 2929$ and $e = 13$ are made public.

Now suppose Elias is an online customer who wishes to send securely to Amira the number $m = 888$. He first computes

$$888^{13} \bmod 2929 = 2705$$

then sends her this number $s = 2705$. Upon receiving s, Amira computes

$$2705^{1077} \bmod 2929 = 888$$

which is the correct intended number (message) from Elias.

Exercise 1.50. In this mini RSA exercise, Amira uses $n = 391$ and $e = 5$.

a) Elias is to give her the message $m = 234$. What is the value of s which he sends to Amira?

b) Find p and q using trial division.

c) Find Amira's secret key b and verify that $s^b \bmod 391 = 234$.

d) Another time Amira receives $s = 319$. Discover the intended message m and cross-check that $m^5 \bmod 391 = 319$.

Exercise* 1.51. The statement $a^{p-1} \bmod p = 1$ in Fermat's little theorem relies on the assumption that p does not divide a. The RSA algorithm, however, assumes the theorem without knowing whether p or q divides m. In theory, the probability of such occurrence is extremely small in view of the abundance of primes their size. Nevertheless, please modify the RSA argument to confirm that (1.7) remains valid even if p or q divides m.

1.4.5   Recognizing Large Composites

We have seen that with trial division we can factor any integer, at least theoretically, or prove that it is prime. There are times, as in RSA, when we need to distinguish large primes from composites. We will see two algorithms which can be used to identify large composites without resorting to factorization. While they may not work for *all* composites, these algorithms are still far superior than trial division in time efficiency.

The first such algorithm is based on Fermat's little theorem. The statement of Theorem 1.12 holds whenever p is prime; so if it fails for some integer $p = n$, whose primality is to be determined, then we may safely conclude that n is composite.

Example. Given $n = 989$. Choosing $a = 2$, we use the successive squaring algorithm to discover that $2^{988} \bmod 989 = 213 \neq 1$, a result which would violate Fermat's little theorem if 989 were prime. Hence, we conclude that 989 is composite.

Question. Have you wondered why Fermat's theorem is called *little*?

Fermat's little theorem, however, is not designed to recognize a prime number. What this means is, if $a^{n-1} \bmod n = 1$, we are not allowed to hastily conclude that n is a prime. See, for instance, that $2^{340} \bmod 341 = 1$, and yet 341 is genuinely composite, as $341 = 11 \times 31$. What we can do in such a case is perhaps try another value of a, e.g., $3^{340} \bmod 341 = 56 \neq 1$, which confirms that 341 is indeed composite.

Exercise 1.52. Which ones of the following numbers are recognized as composites using Fermat's little theorem with base $a = 2$ and/or $a = 3$?

a) 561
b) 779
c) 1013
d) 1387

Definition. Suppose that $a^{n-1} \bmod n = 1$ for some integer $a \geq 2$ and some odd number n. If the number n is composite, then we call n a *Fermat pseudoprime* base a.

As we have just seen, the number 341 is a Femat pseudoprime base 2, but not base 3. The worst kind of a Fermat pseudoprime is when $a^{n-1} \bmod n = 1$ holds for many values of a. In fact, the Carmichael numbers n, defined next, are Fermat pseudoprimes to all bases a as long as $\gcd(a, n) = 1$.

Definition. A composite n is called a *Carmichael number* when n factors into distinct primes such that for each prime factor p, the number $p - 1$ divides $n - 1$.

For example, 561 is a Carmichael number because $561 = 3 \times 11 \times 17$, all distinct primes, and 560 is divisible by 2, by 10, and by 16. In fact, 561 is actually the smallest Carmichael number.

Exercise 1.53. Use trial division to factor each number, then verify that the composite is a Carmichael number.
a) 1729
b) 2465
c) 6601
d) 8911

Exercise 1.54. Find two examples of a prime $p < 100$ such that the number $n = 7 \times 31 \times p$ is Carmichael.

Exercise* 1.55. Show why a Carmichael number must be odd.

Theorem 1.13. Suppose that $\gcd(a, n) = 1$. If n is a Carmichael number, then n is a Fermat pseudoprime base a.

Proof. We will demonstrate the claim for $n = 561$ in a structural way which readily applies to all Carmichael numbers n in general.

Let $\gcd(a, 561) = 1$, so a is not a multiple of 3, 11, or 17. By Fermat's little theorem we have $a^{p-1} \bmod p = 1$ for each $p = 3$, 11, and 17. Since

$$a^{560} = (a^2)^{280} = (a^{10})^{56} = (a^{16})^{35}$$

we see by Theorem 1.11 that $a^{560} \bmod p = 1$ for each $p = 3$, 11, and 17. It follows that 3, 11, and 17 are all prime factors of the number $a^{560} - 1$. And as $3 \times 11 \times 17 = 561$, we conclude that $a^{560} \bmod 561 = 1$. $\qquad \triangledown$

Although rare, it has been discovered that Carmichael numbers are infinitely many. If the job is to catch composites, Fermat's little theorem is therefore rather weak at it. A stronger compositeness test is based on the following observation.

Theorem 1.14. If p is a prime and $x^2 \bmod p = 1$ for some integer x, then either $x \bmod p = 1$ or $x \bmod p = p - 1$.

Proof. We have p dividing $x^2 - 1 = (x + 1)(x - 1)$. By Theorem 1.7, either p divides $x + 1$ or $x - 1$; the former implies $x \bmod p = p - 1$ and the latter $x \bmod p = 1$.   ▽

Theorem 1.14 may not hold for composites, e.g., $5^2 \bmod 12 = 1$, where neither $5 \bmod 12 = 1$ nor $5 \bmod 12 = 11$ is true. In fact, this is the idea: if $a^{n-1} \bmod n = 1$ and we suspect that n might be a pseudoprime, we will look at $a^{(n-1)/2} \bmod n$. If this last quantity is neither 1 nor $n - 1$ then, failing the theorem, n must be a composite. The full algorithm is given as the next compositeness test.

Theorem 1.15 (Miller-Rabin Test)**.** Let n be an odd integer whose primality is to be determined, and fix a base number a such that $\gcd(a, n) = 1$. Write $n - 1 = 2^e \times d$ where d is odd, and consider the sequence given by

$$a^d \bmod n, \ a^{2d} \bmod n, \ a^{4d} \bmod n, \ a^{8d} \bmod n, \ \dots \ , \ a^{n-1} \bmod n$$

If a term equals 1 and is preceded by neither 1 nor $n-1$, then n is composite.

Proof. Each successive term is obtained by squaring the previous one, hence by Theorem 1.14, a 1 must be preceded by 1 or $n - 1$, if n be prime.   ▽

Note that the sequence consists of $e + 1$ numbers in all, the last term being $a^{2^e \times d} \bmod n$. Moreover, if this last term is not 1, then n is composite, but that is Fermat's little theorem.

Example. We try the Carmichael number 561 for Miller-Rabin test with $a = 2$. Since $560 = 2^4 \times 35$, there are 5 terms in our sequence:

$$2^{35} \bmod 561, \ 2^{70} \bmod 561, \ 2^{140} \bmod 561, \ 2^{280} \bmod 561, \ 2^{560} \bmod 561$$

Using successive squaring algorithm, this sequence turns out to be

$$263, \ 166, \ 67, \ 1, \ 1$$

Note the term 1 preceded by 67, so we conclude that 561 is composite.

Exercise 1.56. Test the Carmichael numbers given in Exercise 1.53 using Theorem 1.15. Which ones are recognized as composites?

Still, Miller-Rabin test may miss some composites which go undetected by Theorem 1.15. We call such odd composites *strong pseudoprimes* base a. The smallest strong pseudoprime base 2 is $2047 = 23 \times 89$. You may verify, with $2046 = 2 \times 1023$, that the two terms in the sequence are just 1 and 1.

Exercise 1.57. The following composites are all Fermat pseudoprimes base 2. Which ones are also strong pseudoprimes base 2?

a) 1105
b) 2821
c) 3277
d) 4033

Exercise* 1.58. Explain why every strong pseudoprime is necessarily a Fermat pseudoprime, to the same base.

As a final remark, although strong pseudoprimes do exist, Theorem 1.15 can nevertheless be used to recognize primes within certain bounds. It has been tested, for instance, that there are no strong pseudoprimes less than 2 trillion to the bases 2, 3, 5, 7, and 11 simultaneously. Hence, within this huge interval, a number n which "passes" Miller-Rabin test to these five bases must be a genuine prime.

Books to Read

1. D. M. Bressoud, *Factorization and Primality Testing,* Springer 1989.

2. S. C. Coutinho, *The Mathematics of Ciphers: Number Theory and RSA Cryptography,* A K Peters 1999.

3. O. Ore, *Number Theory and Its History,* 1948, Dover Publications 1988.

4. W. Trappe and L. C. Washington, *Introduction to Cryptography with Coding Theory,* Second Edition, Prentice Hall 2005.

Chapter 2

Topics in Logic and Proofs

Some mathematical statements carry a logical value of being true or false, while some do not. For example, the statement "$4 + 5 = 9$" is true, whereas the statement "2 is odd" is false. However, a statement like "$x^2 - 2x = 15$" is neither true nor false, until further information is given concerning x. We now make the following definition.

Definition. By a *truth value* we mean a logical value of true or false. A statement which possesses a truth value is called a *proposition*.

Technically, of course, a proposition can be stated in any language, not necessarily mathematical; the only requirement is that the statement must be quantifiable as being true or false.

This leads to the algebra of *boolean logic*, in which we are dealing with entities whose values can either be 0 (false) or 1 (true). In fact, this reminds us of the binary number system and the underlying structure (on/off switches) of computing machines.

2.1 Propositional Logic

As with numbers, we now treat propositions as mathematical quantities which can be operated one on another by a selection of proposition operators, or *logic operators*. The first and simplest operator is analogous to taking the negative of a number.

Definition. Let p denote a proposition. The *negation* of p is the proposition given by the statement "not p" and whose value is opposite that of p. The negation of p can simply be called *not p* and is denoted by $\neg p$.

Example. We give two propositions, one in mathematics and another in English, each with its negation.

$$p: \quad 4+5 = 9 \qquad q: \quad \text{The earth is flat.}$$
$$\neg p: \quad 4+5 \neq 9 \qquad \neg q: \quad \text{The earth is not flat.}$$

Note that each proposition has the opposite truth value from that of its negation; If p is true then $\neg p$ is false, and vice versa.

Question. What would be the value of the proposition $\neg(\neg p)$?

2.1.1 Logic Operators and Truth Tables

A logic operator can be given by a table which displays the output value for every possible combination of the input values. The *truth table* for the negation operator, for instance, is given below.

Table 2.1: Truth table for $\neg p$.

p	$\neg p$
T	F
F	T

A number of logic operators will now be given by their truth tables. In general, the resulting proposition obtained by applying these operators will be called a *compound* proposition.

Definition. Let p and q be two propositions. The *conjunction* $p \wedge q$ and *disjunction* $p \vee q$ yield the compound statements p *and* q, respectively, p *or* q, and whose values are given according to the following table.

Table 2.2: Truth tables for $p \wedge q$ and $p \vee q$.

p	q	$p \wedge q$	$p \vee q$
T	T	T	T
T	F	F	T
F	T	F	T
F	F	F	F

Example. Suppose $p: 4+5 = 9$ and $q: 2$ is odd. Write the statement and find the value of the compound proposition (a) $\neg p \wedge q$ (b) $p \vee \neg q$.

Solution. Note that p has value true and q false. The first statement is false $(F \wedge F)$ and the second true $(T \vee T)$, and they are given by

a) $\neg p \wedge q : 4 + 5 \neq 9$ and 2 is odd
b) $p \vee \neg q : 4 + 5 = 9$ or 2 is not odd

Exercise 2.1. Suppose p is true and q is false. Determine true or false for each compound proposition below.

a) $\neg p \vee \neg q$
b) $(p \wedge \neg q) \vee \neg p$
c) $(p \wedge q) \vee (\neg p \wedge \neg q)$
d) $(\neg p \vee (q \vee p)) \wedge (p \wedge q)$

Example. Construct a truth table to determine the possible output values of the compound proposition given by $(p \vee \neg q) \wedge (\neg p \vee q)$.

Solution. There are four possible rows. We show the intermediate steps according to the order in which the logic operations apply, as follows.

p	q	$\neg p$	$\neg q$	$p \vee \neg q$	$\neg p \vee q$	$(p \vee \neg q) \wedge (\neg p \vee q)$
T	T	F	F	T	T	T
T	F	F	T	T	F	F
F	T	T	F	F	T	F
F	F	T	T	T	T	T

Exercise 2.2. Construct the truth table for each given compound proposition.

a) $\neg(\neg p \wedge \neg q)$
b) $\neg p \vee (p \wedge \neg q)$
c) $(p \wedge \neg q) \vee (\neg p \vee q)$
d) $(p \vee q) \wedge (\neg p \wedge \neg q)$

Definition. The *implication* $p \to q$ yields a compound proposition whose truth value is given in Table 2.3. The statement $p \to q$ is read *if p then q*, or sometimes, *p implies q*. Implication is also called the *if-then* operator.

Table 2.3: Truth table for $p \to q$.

p	q	$p \to q$
T	T	T
T	F	F
F	T	T
F	F	T

Question. Do $p \to q$ and $q \to p$ always have the same truth value?

Exercise 2.3. Suppose $p : 4 + 5 = 9$ and $q : 2$ is odd. Write the statement and determine the value of each compound proposition below.

a) $p \rightarrow q$
b) $q \rightarrow p$
c) $\neg p \rightarrow q$
d) $\neg q \rightarrow \neg p$

Example. Construct the truth table for the proposition $(p \rightarrow q) \rightarrow r$.

Solution. This is the first time we see a compound proposition involving three propositional variables. The first three columns of the next table show the standard ordering for the eight possibilities of the values of (p, q, r).

p	q	r	$p \rightarrow q$	$(p \rightarrow q) \rightarrow r$
T	T	T	T	T
T	T	F	T	F
T	F	T	F	T
T	F	F	F	T
F	T	T	T	T
F	T	F	T	F
F	F	T	T	T
F	F	F	T	F

Question. How many rows are in the truth table, if there are four variables, p, q, r, s, in the compound proposition?

Exercise 2.4. Construct the truth table for each given compound proposition.

a) $\neg q \rightarrow \neg p$
b) $(p \wedge q) \rightarrow (p \vee q)$
c) $(p \vee q) \rightarrow r$
d) $(\neg p \rightarrow q) \wedge (\neg p \rightarrow r)$

Definition. The compound propositions $p \leftrightarrow q$ (read p *if and only if* q, or p *iff* q) and $p \oplus q$ (read p *exclusive or* q, or p *xor* q) are given by their truth tables, respectively, next.

Table 2.4: Truth tables for $p \leftrightarrow q$ and $p \oplus q$.

p	q	$p \leftrightarrow q$	$p \oplus q$
T	T	T	F
T	F	F	T
F	T	F	T
F	F	T	F

In the English language, the exclusive or is often translated p *or* q *but not both,* since the table shows that $p \oplus q$ is true when exactly one of them is true, not both. Moreover, a proposition of the form $p \leftrightarrow q$ is called a *biconditional* statement, and is used to connect two statements whose truth values are the same, i.e., p is true if q is true, and p is false if q is false.

Note that these two compound propositions have opposite values for each pair (p, q). To help remember, $p \leftrightarrow q$ is true exclusively when p and q have identical values, whereas $p \oplus q$ is true exactly when p and q have unequal truth values.

Exercise 2.5. Suppose that we have the following propositions.

$p :$ It is hot today.
$q :$ It is windy today.
$r :$ It will rain tomorrow.

Translate the following sentences using the variables p, q, r, and the appropriate logic operators.

a) If today is hot and windy, then it will rain tomorrow.
b) Tomorrow will rain if and only if today is not windy.
c) Either today is hot or tomorrow will rain, but not both.
d) If today is neither hot nor windy, then it will not rain tomorrow.

Exercise 2.6. Construct the truth table for each given compound proposition.

a) $(p \leftrightarrow q) \wedge (p \oplus q)$
b) $(p \leftrightarrow \neg q) \rightarrow (\neg p \oplus \neg q)$
c) $(p \oplus q) \oplus r$
d) $[(\neg p \wedge q) \vee (\neg r \rightarrow p)] \leftrightarrow (r \oplus \neg q)$

2.1.2   Tautology and Contradiction

Consider the compound proposition $(p \wedge q) \rightarrow p$, whose truth table, displayed below, happens to show all true values. This is an example of a tautology.

p	q	$p \wedge q$	$(p \wedge q) \rightarrow p$
T	T	T	T
T	F	F	T
F	T	F	T
F	F	F	T

Definition. A *tautology* is a compound proposition whose truth table consists of all true values.

Test 2.7. Which one of the following propositions is a tautology?

a) $p \wedge \neg p$
b) $p \rightarrow \neg p$
c) $p \leftrightarrow \neg p$
d) $p \oplus \neg p$

Definition. The counterpart of a tautology is a *contradiction*, i.e., a compound proposition which shows all false values in the truth table. Incidently, a compound proposition whose table contains a mix of true and false, like most that we have seen thus far, is called a *contingency*.

A ready example of a contradiction is given by $\neg((p \wedge q) \rightarrow p)$. (Why?) Quite expectedly, both tautologies and contradictions are rather rare occurrences and, given a random compound proposition, chances are it is a contingency more likely than it is otherwise.

Test 2.8. Which one of the following propositions is a contingency?

a) $p \wedge p$
b) $p \rightarrow p$
c) $p \leftrightarrow p$
d) $p \oplus p$

Exercise 2.9. Identify each compound proposition as a tautology, contradiction, or contingency.

a) $p \rightarrow (p \vee q)$
b) $(p \rightarrow q) \rightarrow q$
c) $(p \leftrightarrow q) \wedge (p \oplus q)$
d) $(p \leftrightarrow \neg q) \rightarrow (\neg p \oplus \neg q)$

2.1.3 Logical Arguments

In everyday's English we sometimes use the word *argument* to mean a reasoning, or a process of thought based on a set of assumptions from which we derive a conclusion. We consider an argument valid when the conclusion can be proved to be a logically sound consequence of the assumptions, disregarding the truth values of the assumptions themselves. For example, if we assume that all odd numbers are primes, then we may conclude that 15 is prime. Here, the argument is valid—because 15 is an odd number—even though the conclusion is actually false.

Definition. By a *premise* we mean a proposition whose value is assumed true. An *argument* is a set of premises $P_1, P_2, \ldots, P_k$, together with another proposition Q which serves as the *conclusion*. The argument is said to be *valid* when the proposition $(P_1 \wedge P_2 \wedge \cdots \wedge P_k) \rightarrow Q$ is a tautology.

Example. Assume the following two premises.

P_1 :   Tomorrow is not Friday.
P_2 :   If today is not Sunday then tomorrow is Friday.

Therefore, we claim the following conclusion.

Q :   Today is Sunday.

Is the above argument valid?

Solution. Let us fix the following propositions.

p :   Today is Sunday.
q :   Tomorrow is Friday.

The two premises and the conclusion are then represented by, respectively,

P_1 : $\neg q$
P_2 : $\neg p \to q$
Q : p

We need now study the truth table of the compound proposition

$$(P_1 \wedge P_2) \to Q : (\neg q \wedge (\neg p \to q)) \to p$$

given below.

p	q	$\neg p$	$\neg q$	$\neg p \to q$	$\neg q \wedge (\neg p \to q)$	$(\neg q \wedge (\neg p \to q)) \to p$
T	T	F	F	T	F	T
T	F	F	T	T	T	T
F	T	T	F	T	F	T
F	F	T	T	F	F	T

The table shows that $(P_1 \wedge P_2) \to Q$ is indeed a tautology, establishing the validity of the argument.

Example. Assume that every even number is composite. Can we conclude that all odd numbers are prime?

Solution. Let p denote the statement "n is even" and q the statement "n is composite." Note that the premise is given by $p \to q$, and the conclusion $\neg p \to \neg q$. We look at the truth table, and find a contingency. Hence, the argument is not valid.

p	q	$p \to q$	$\neg p$	$\neg q$	$\neg p \to \neg q$	$(p \to q) \to (\neg p \to \neg q)$
T	T	T	F	F	T	T
T	F	F	F	T	T	T
F	T	T	T	F	F	F
F	F	T	T	T	T	T

Exercise 2.10. Determine the validity of each given argument.

a) Premises: Today is not Sunday. Today is Sunday if and only if tomorrow is Tuesday. Conclusion: Tomorrow is not Tuesday.
b) Premises: If you like Discrete Mathematics, you will like Calculus. You like neither Discrete Mathematics nor Calculus. Conclusion: If you like Calculus, you will like Discrete Mathematics.
c) Premises: If n is even then n is composite. If n is prime then $2n + 1$ is also prime. Conclusion: If $2n + 1$ is composite then either n is prime or odd.
d) Premises: Either prime numbers are infinitely many or composites are, but not both. There are infinitely many primes. If composites are finitely many, so are even numbers. Conclusion: Both composites and even numbers are finitely many.

The following theorem lists a few conditional statements which are rather well-known tautologies. They can be used as models for a valid argument, and are sometimes refered to as *rules of inference*.

Theorem 2.1. Each of the following propositions is a tautology.

$$(p \wedge q) \to p \qquad\qquad p \to (p \vee q)$$
$$\neg p \to (p \to q) \qquad\qquad \neg (p \to q) \to p$$
$$(p \wedge (p \to q)) \to q \qquad\qquad (\neg q \wedge (p \to q)) \to \neg p$$
$$(\neg p \wedge (p \vee q)) \to q \qquad\qquad ((p \to q) \wedge (q \to r)) \to (p \to r)$$

Proof. We leave it to you to explore the truth tables for all the propositions stated above. ▽

Test 2.11. Let us agree on three premises: The earth is flat if and only if the moon is. If the earth is not flat, then neither is the sun. But if the sun is flat, so must the moon be. Which one of the following conclusions makes an invalid argument?

a) If the sun is flat, so are the moon and the earth.
b) If the moon is flat, so are the earth and the sun.
c) If the earth is not flat, neither are the sun and the moon.
d) All of the above conclusions are valid.

2.1.4 Logical Equivalence

Sometimes it may well be the case that two compound propositions have look-alike truth tables. Can you see, for instance, why the table for $p \oplus q$ is no different that that for $\neg (p \leftrightarrow q)$? Such a relation between two propositions is an important concept and shall be given a special name.

Definition. Two propositions are called *equivalent* to each other if their truth tables are identical. We employ the symbol $\equiv$ to denote this relation. Hence, for example, we have $\neg(p \leftrightarrow q) \equiv p \oplus q$.

Example. Prove the relation $p \to q \equiv \neg p \vee q$.

Solution. We have to create the two tables and arrive at the same results. To save some space, we will juxtapose the two tables into one as follows.

p	q	$p \to q$	$\neg p$	$\neg p \vee q$
T	T	T	F	T
T	F	F	F	F
F	T	T	T	T
F	F	T	T	T

Note that both final columns show identical entries, justifying the equivalence between the two propositions.

Exercise 2.12. Verify the equivalence in each of the following statements.

a) $\neg p \wedge q \equiv \neg(p \vee \neg q)$
b) $p \leftrightarrow q \equiv (p \to q) \wedge (q \to p)$
c) $p \wedge (q \vee r) \equiv (p \wedge q) \vee (p \wedge r)$
d) $p \to (q \to r) \equiv q \to (p \to r)$

Exercise 2.13. Prove that all of the following compound propositions are equivalent one to another.

a) $p \to \neg q$
b) $q \to \neg p$
c) $\neg p \vee \neg q$
d) $\neg(p \wedge q)$

Test 2.14. Which one of the following is *not* equivalent to $p \oplus q$?

a) $p \leftrightarrow \neg q$
b) $\neg p \leftrightarrow q$
c) $\neg p \leftrightarrow \neg q$
d) $\neg p \oplus \neg q$

2.1.5   Implication and Its Contrapositive

Definition. Given an implication of the form $p \to q$, we define its *contrapositive* to be the proposition given by $\neg q \to \neg p$.

Theorem 2.2. An implication is always equivalent to its contrapositive.

Proof. We show the equivalence $p \to q \equiv \neg q \to \neg p$ by simply producing their respective tables below. ▽

p	q	$p \to q$	$\neg q$	$\neg p$	$\neg q \to \neg p$
T	T	T	F	F	T
T	F	F	T	F	F
F	T	T	F	T	T
F	F	T	T	T	T

Example. The English sentence "If today is Friday, then tomorrow is Saturday" is in the form of an implication. Convert the sentence using its contrapositive.

Solution. Let p represent the statement "Today is Friday" and q "Tomorrow is Saturday." The sentence we wish to convert is represented by $p \to q$. The contrapositive $\neg q \to \neg p$ says "If tomorrow is *not* Saturday, then today is *not* Friday." Can you see that the two sentences are equivalent in their meaning? They say the same thing in two different ways.

Exercise 2.15. Rewrite each statement using its contrapositive.

a) Today is Friday, if tomorrow is not Sunday.
b) When x is not an integer, neither is x^2.
c) An even number is never a prime.
d) Every mathematician is wealthy.

Question. What is the contrapositive of René Descartes's famous philosophical quote "I think, therefore I am" (Cogito ergo sum)?

Test 2.16. Which one of the following is equivalent to $\neg q \to p$?

a) $p \to \neg q$
b) $\neg p \to q$
c) $\neg p \to \neg q$
d) $q \to \neg p$

Definition. The *converse* of an implication $p \to q$ is the proposition $q \to p$.

For example, the converse of "If today is Friday, then tomorrow is Saturday" is given by "If tomorrow is Saturday, then today is Friday."

Exercise 2.17. Write the converse of each statement given in Exercise 2.15.

Question. For which truth values of (p, q) can we have $p \to q \equiv q \to p$?

2.1.6   Common Equivalence Rules

Equivalence between two compound propositions, in a sense, allows the substitution of one by the other without altering its truth value. The following list contains some of the most common equivalence rules which may come in handy in problem solving.

Theorem 2.3. The following equivalence rules hold.

1. $p \wedge q \equiv q \wedge p$
   $p \vee q \equiv q \vee p$

2. $p \wedge (q \wedge r) \equiv (p \wedge q) \wedge r$
   $p \vee (q \vee r) \equiv (p \vee q) \vee r$

3. $p \wedge (q \vee r) \equiv (p \wedge q) \vee (p \wedge r)$
   $p \vee (q \wedge r) \equiv (p \vee q) \wedge (p \vee r)$

4. $\neg(\neg p) \equiv p$
   $\neg(p \wedge q) \equiv \neg p \vee \neg q$
   $\neg(p \vee q) \equiv \neg p \wedge \neg q$

5. $p \rightarrow q \equiv \neg p \vee q$
   $p \leftrightarrow q \equiv (p \rightarrow q) \wedge (q \rightarrow p)$
   $p \oplus q \equiv \neg(p \leftrightarrow q)$

Proof. These results can all be established with the help of truth tables. We omit the proof as an easy exercise.   ▽

Example. Prove $p \rightarrow (q \rightarrow r) \equiv q \rightarrow (p \rightarrow r)$ by applying the above rules.

Solution. A series of substitutions takes place as follows.

$$
\begin{array}{ll}
p \rightarrow (q \rightarrow r) \equiv p \rightarrow (\neg q \vee r) & \text{(Rule 5a)} \\
\equiv \neg p \vee (\neg q \vee r) & \text{(Rule 5a)} \\
\equiv (\neg p \vee \neg q) \vee r & \text{(Rule 2b)} \\
\equiv (\neg q \vee \neg p) \vee r & \text{(Rule 1b)} \\
\equiv \neg q \vee (\neg p \vee r) & \text{(Rule 2b)} \\
\equiv q \rightarrow (\neg p \vee r) & \text{(Rule 5a)} \\
\equiv q \rightarrow (p \rightarrow r) & \text{(Rule 5a)}
\end{array}
$$

Exercise 2.18. Prove by applying equivalence rules.
a) $\neg(p \rightarrow q) \equiv p \wedge \neg q$
b) $p \rightarrow q \equiv \neg q \rightarrow \neg p$
c) $p \rightarrow (q \wedge r) \equiv (p \rightarrow q) \wedge (p \rightarrow r)$
d) $p \oplus q \equiv (p \wedge \neg q) \vee (q \wedge \neg p)$

Exercise 2.19. Some of the following statements are false. Determine the validity of each one, using any method you prefer.
a) $p \rightarrow (q \rightarrow r) \equiv (p \rightarrow q) \rightarrow r$
b) $p \rightarrow (q \vee r) \equiv (p \rightarrow q) \vee (p \rightarrow r)$
c) $p \vee (q \oplus r) \equiv (p \vee q) \oplus (p \vee r)$
d) $\neg(p \oplus q) \equiv \neg p \leftrightarrow \neg q$

2.1.7 CNF and DNF

We have seen that a logic operator, such as $p \to q$, is defined by its truth table. In other words, a different table gives a different logic operator.

Question. How many different logic operators involving p and q are possible?

There is no doubt, however, that some of these operators are actually interchangeable via certain equivalence rules and hence not all of them are quite necessary to have. The next theorem, in fact, claims that every compound proposition can eventually be written using only conjunctions, disjunctions, and negations!

Definition. By a *conjunctive normal form*, or *CNF*, we mean a series of conjunctions operating on any number of compound propositions, each of which is the disjunctions of propositional variables or their negations. An example of a CNF involving three variables p, q, r, is

$$(p \vee \neg q \vee r) \wedge (p \vee \neg q \vee \neg r) \wedge (\neg p \vee q \vee r) \wedge (\neg p \vee \neg q \vee r)$$

Similarly, we define a *disjunctive normal form*, or *DNF*, by reversing the roles of disjunction and conjunction in the CNF definition. Thus an example of a DNF with four variables could be

$$(p \wedge q \wedge \neg r \wedge s) \vee (p \wedge \neg q \wedge r \wedge \neg s) \vee (\neg p \wedge \neg q \wedge \neg r \wedge s)$$

Note that there is no limit to the number of variables nor to the number of brackets involved. A *normal form* refers to either a CNF or a DNF.

Theorem 2.4. Every compound proposition is equivalent to a CNF and to a DNF.

We will not formally prove this theorem. Instead, the next two examples will illustrate how and why the theorem works.

Example. Convert the given proposition below to a CNF and to a DNF.

$$(p \leftrightarrow q) \to (p \oplus q)$$

Solution. The first step is to construct the truth table, labeling the rows 1 to 4:

	p	q	$p \leftrightarrow q$	$p \oplus q$	$(p \leftrightarrow q) \to (p \oplus q)$
1	T	T	T	F	F
2	T	F	F	T	T
3	F	T	F	T	T
4	F	F	T	F	F

The rows with *false* values, i.e., rows 1 and 4, will give us a CNF of two brackets, where the two variables p and q are negated if necessary, such that both values are *false* at each corresponding row:

$$(p \leftrightarrow q) \rightarrow (p \oplus q) \equiv (\text{row } 1) \wedge (\text{row } 4)$$
$$\equiv (\neg p \vee \neg q) \wedge (p \vee q) \qquad (\text{CNF})$$

But why is this algorithm correct? To justify the equivalence, simply verify that this CNF gives F T T F in its truth table. Without actually displaying the table, note that row 1 will be false because the first bracket is false (disjunction of two false propositions). In the same way, row 4 will be false since the second bracket is false. For the unselected rows 2 and 3, each bracket will be true since one of the two variables must be true (by design, both can be false simultaneously *only* in rows 1 or 4) so that the resulting conjunction is also true.

Similarly for DNF we select the *true* rows, and this time we want the variables to be *true* as well:

$$(p \leftrightarrow q) \rightarrow (p \oplus q) \equiv (\text{row } 2) \vee (\text{row } 3)$$
$$\equiv (p \wedge \neg q) \vee (\neg p \wedge q) \qquad (\text{DNF})$$

Example. For a second illustration, suppose that a compound proposition X of three variables generates a truth table as given below. Convert the proposition X to a CNF and to a DNF.

	p	q	r	$\cdots$	X
1	T	T	T	$\cdots$	T
2	T	T	F	$\cdots$	T
3	T	F	T	$\cdots$	F
4	T	F	F	$\cdots$	F
5	F	T	T	$\cdots$	T
6	F	T	F	$\cdots$	T
7	F	F	T	$\cdots$	T
8	F	F	F	$\cdots$	F

Solution. For the CNF we take the false rows 3, 4, and 8. The variables p, q, r will not be negated if already false:

$$X \equiv (\text{row } 3) \wedge (\text{row } 4) \wedge (\text{row } 8)$$
$$\equiv (\neg p \vee q \vee \neg r) \wedge (\neg p \vee q \vee r) \wedge (p \vee q \vee r) \qquad (\text{CNF})$$

We leave it to you to write down the DNF based on the remaining five rows.

Exercise 2.20. Convert to a CNF and to a DNF.

a) $\neg(p \wedge q) \rightarrow p$
b) $(p \oplus \neg q) \leftrightarrow (\neg p \vee q)$
c) $(p \rightarrow q) \rightarrow r$
d) $((p \wedge q) \rightarrow r) \oplus (\neg p \vee (q \leftrightarrow r))$

Question. Given a CNF, how can we quickly convert it to a DNF, or vice versa? Do we really need to construct the truth table?

Test 2.21. Convert the CNF $(p \vee \neg q) \wedge (\neg p \vee \neg q)$ to a DNF.

a) $(\neg p \wedge q) \vee (p \wedge q)$
b) $(\neg p \wedge q) \vee (\neg p \wedge \neg q)$
c) $(p \wedge \neg q) \vee (p \wedge q)$
d) $(p \wedge \neg q) \vee (\neg p \wedge \neg q)$

2.2 Introduction to Sets

The word set loosely means a collection or group of objects. In the present context, however, we shall define this term more restrictively as follows.

Definition. A *set* is a collection of any objects in which ordering and repetition of its members are ignored. The objects which make up the members of the set are called the *elements* of the set.

We assume the common notation of a set using a pair of braces and of membership using $\in$, e.g., $S = \{a, e, i\}$ represents the set S whose elements are the first three vowels of the English alphabet: a, e, i. Of this example we may state that $a \in S$, whereas $u \notin S$. Note that the sets $\{a, e, i\}$ and $\{i, e, a\}$, as well as $\{a, i, a, e, a, e\}$ are to be treated as identical by our definition. More precisely,

Definition. Two sets A and B are identical, in which case we write $A = B$, when the following proposition holds for every element x.

$$x \in A \leftrightarrow x \in B$$

Having agreed on this, henceforth we shall use the following notation for some very common number sets.

$\mathbb{Z}$ the set of integers

$\mathbb{N}$ the set of positive integers

$\mathbb{Q}$ the set of rational numbers

$\mathbb{R}$ the set of real numbers

It is convenient, for instance, to simply write $\mathbb{N}$, which invariably refers to the set $\{1, 2, 3, \ldots\}$ throughout this text, rather than the phrase "the set of positive integers" or "the set of natural numbers."

If A is a given set and $P(x)$ is a statement whose truth value shall be determined by the variable x, we may at times have a set S whose elements are given implicitly in the following *set-builder notation*.

$$S = \{x \in A \mid P(x)\}$$

This is to mean that $a \in S$ if and only if $a \in A$ and for which $P(a)$ is true. If, however, the property that $x \in A$ is already implied in the statement $P(x)$, then we may just write $S = \{x \mid P(x)\}$. Hence, for example, the set of even numbers can be written as $\{x \in \mathbb{Z} \mid x \bmod 2 = 0\}$. Similarly, we have $\{x \in \mathbb{Z} \mid x > 0\} = \mathbb{N}$.

Question. What are the elements in the set $\{x \in \mathbb{R} \mid x^2 + 1 = 0\}$?

Exercise 2.22. Describe each set given below; they are all familiar number sets.

a) $\{x \in \mathbb{Z} \mid x \bmod 2 = 1\}$
b) $\{x \in \mathbb{Q} \mid x/2 \in \mathbb{Z}\}$
c) $\{a/b \mid a \in \mathbb{Z} \wedge b \in \mathbb{N}\}$
d) $\{x \in \mathbb{R} \mid nx \notin \mathbb{Z} \text{ for all } n \in \mathbb{Z}\}$

A set is allowed to be empty, that is, to have no elements whatsoever in it. We reserve the notation $\emptyset$ (read *phi*) to always denote the empty set, as far as set theory is concerned. Thus, $\emptyset = \{\ \}$.

Question. Is it true that $\{\emptyset\} = \emptyset$?

Test 2.23. Which one of the following sets is *not* empty?

a) $\{x \in \mathbb{Z} \mid x \bmod 2 = 2\}$
b) $\{x \in \mathbb{Q} \mid x^2 = 2\}$
c) $\{x \in \mathbb{R} \mid x^2 = -1\}$
d) $\{x \in \mathbb{R} \mid x^2 < x\}$

2.2.1   Set Operators

Two sets can be operated on and yield a new set, in ways which resemble the logical operations on two propositions that we have learned in the previous section. We introduce the first four of such operators in the next definition.

Definition. Let A and B be any two sets. The operations *union* $A \cup B$, *intersection* $A \cap B$, *difference* $A - B$, and *symmetric difference* $A \oplus B$ are given by their set-builder notation, respectively, as follows.

$$A \cup B = \{x \mid x \in A \vee x \in B\}$$
$$A \cap B = \{x \mid x \in A \wedge x \in B\}$$
$$A - B = \{x \mid x \in A \wedge x \notin B\}$$
$$A \oplus B = \{x \mid x \in A \oplus x \in B\}$$

Note that the set operator symmetric difference in $A \oplus B$ is given by the logical operator exclusive or, appearing in the statement $x \in A \oplus x \in B$— the same notation, and essentially the same operation, with two different names, one applying for sets and the other for propositions.

Example. Let $A = \{1, 3, 5, 7\}, B = \{0, 1, 2, 3\}$, and $C = \{0, 2\}$. Determine the output of each set operation given below.

a) $A \cup B, A \cup C, B \cup C$
b) $A \cap B, A \cap C, B \cap C$
c) $A - B, A - C, B - C$
d) $A \oplus B, A \oplus C, B \oplus C$

Solution. We refer to the four definitions given above.

a) By definition, $x \in A \cup B$ if and only if $x \in \{1, 3, 5, 7\}$ or $x \in \{0, 1, 2, 3\}$. For this to hold, x can be any one of the elements in either set. Hence, $A \cup B = \{0, 1, 2, 3, 5, 7\}$. Similarly, $A \cup C = \{0, 1, 2, 3, 5, 7\}$ and $B \cup C = \{0, 1, 2, 3\} = B$.
b) Since $x \in A \wedge x \in B$ is true if only if x is a common element of A and B, then we have $A \cap B = \{1, 3\}$. Similarly, $A \cap C = \emptyset$ and $B \cap C = C$.
c) Of the elements in $A = \{1, 3, 5, 7\}$, only 5 and 7 do not belong to B. Hence, $A - B = \{5, 7\}$. Similarly, $A - C = A$ and $B - C = \{1, 3\}$.
d) Note that $x \in \{1, 3, 5, 7\} \oplus x \in \{0, 1, 2, 3\}$ is true exactly when x belongs to one, but not both, of the two sets. So we have $A \oplus B = \{0, 2, 5, 7\}$. Similarly, $A \oplus C = \{0, 1, 2, 3, 5, 7\}$ and $B \oplus C = \{1, 3\}$.

Question. Is it true that $A \cup B = B \cup A$? What about $\cap$, or $-$, or $\oplus$?

Exercise 2.24. Let $A = \{x \in \mathbb{Z} \mid 3 < x < 9\}, B = \{0, 2, 4, 6, 8\}$, and $C = \{x \in A \mid x \bmod 2 = 1\}$. In the following set operations, write out the elements of each resulting set.

a) $(A - B) \cup (A \cap C)$
b) $(A \cap B) \oplus (B \cup C)$
c) $(C \oplus B) - B$
d) $C - (A \oplus C)$

Exercise 2.25. Let A be any given set. Describe the sets given below.

a) $A \cup A$
b) $A \cap A$
c) $A - A$
d) $A \oplus A$

Definition. When $A \cap B = \emptyset$, we say that the two sets A and B are *disjoint*.

Example. Explain why $A \oplus B = A \cup B$ if A and B are disjoint sets.

Solution. We have $x \in A \oplus B$ exactly when x belongs to A or B but not both. The "both" part may be ignored, since $A \cap B$ is empty in this case. Therefore, $A \oplus B = A \cup B$.

Question. Is it possible to have $A \oplus B = A \cup B$ if A and B are *not* disjoint?

Test 2.26. Suppose that A and B are disjoint sets. Which one of the following set identities is true?

a) $A - B = A$
b) $A - B = B$
c) $A \cup B = A$
d) $A \cup B = B$

2.2.2   Venn Diagrams and Set Identities

Venn diagrams are a great tool for describing set operations, in much the same way truth tables are for logical operations. Here the set A is represented by a circle, the region inside of which is where the elements of A reside. If A and B are two sets, then, there are four partitioned regions which we label 1, 2, 3, 4 in the drawing below.

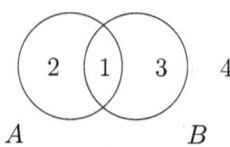

Question. How would you draw the Venn diagram in the special case where A and B are disjoint?

Now if we associate two propositions with these two sets,

$$p: \quad x \in A$$
$$q: \quad x \in B$$

then we see that the four labels respectively correspond to the four rows of (p,q) values in the following truth table.

			$A \cup B$	$A \cap B$	$A - B$	$A \oplus B$
	p	q	$p \vee q$	$p \wedge q$	$p \wedge \neg q$	$p \oplus q$
1	T	T	T	T	F	F
2	T	F	T	F	T	T
3	F	T	T	F	F	T
4	F	F	F	F	F	F

Question. How would you draw the Venn diagram for three sets A, B, C, associated with the three propositions p, q, r, in the truth table?

Note that the table includes the truth values of the four set operations $\cup, \cap, -$, and $\oplus$. These give their Venn diagrams below, showing the regions where each resulting set contains its elements, i.e., where the truth value is true.

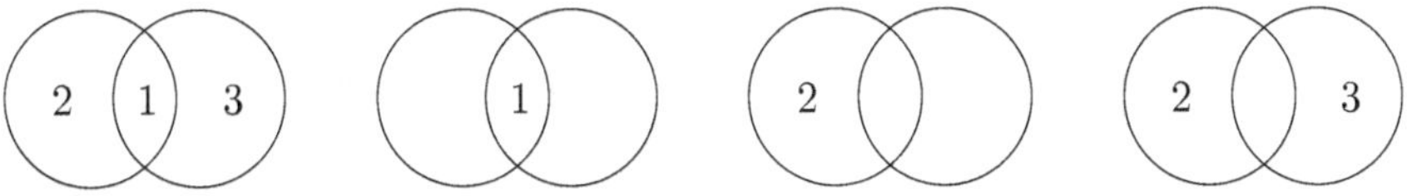

$A \cup B$ $A \cap B$ $A - B$ $A \oplus B$

With Venn diagrams, we are able to give intuitive proofs to certain set identities. It is convincing enough, for instance, to deduce that $(A - B) \cup (A \cap B) = A$ since both consist of regions 1 and 2 in the diagrams. The next theorem is another example.

Theorem 2.5. Let A and B be any two sets. Then

$$A \oplus B = (A - B) \cup (B - A)$$

Proof. The proof is an easy visualization with Venn diagrams. Or, if you prefer truth tables to Venn diagrams, we may let $p : x \in A$ and $q : x \in B$ as before, and note that $(A - B) \cup (B - A)$ is given by the proposition $(p \wedge \neg q) \vee (q \wedge \neg p)$. The truth table below show that $(p \wedge \neg q) \vee (q \wedge \neg p) \equiv p \oplus q$, which defines the set $A \oplus B$. $\triangledown$

p	q	$\neg p$	$\neg q$	$p \wedge \neg q$	$q \wedge \neg p$	$(p \wedge \neg q) \vee (q \wedge \neg p)$
T	T	F	F	F	F	F
T	F	F	T	T	F	T
F	T	T	F	F	T	T
F	F	T	T	F	F	F

Exercise 2.27. Use truth table to establish the set identity

$$A \oplus B = (A \cup B) - (A \cap B)$$

Exercise 2.28. Use Venn diagrams to find the resulting set identical to each one given below.
a) $(A \cap B) \oplus (A - B)$
b) $(A - (A - B)) \oplus B$
c) $(A \cup B) \oplus (A \cap B)$
d) $(A \cup B) \oplus (A \oplus B)$

We have seen that set operations closely resemble logical operations. To make the analogy more complete, we define the negation of a set as follows.

Definition. Let the set A be understood to be part of a larger universal set U. Then by the *complement* of A we mean the set $\neg A = U - A$. For example, if A is the set of even numbers, an appropriate choice of U could be $U = \mathbb{Z}$. In that case, the negation of A is given by $\neg A = \mathbb{Z} - A$, that is, the set of odd numbers.

With that, we are ready to state the following analog of logical equivalence rules for sets.

Theorem 2.6. The following set identities hold.

1. $A \cap B \equiv B \cap A$
    $A \cup B \equiv B \cup A$

4. $\neg(\neg A) \equiv A$
    $\neg(A \cap B) \equiv \neg A \cup \neg B$
    $\neg(A \cup B) \equiv \neg A \cap \neg B$

2. $A \cap (B \cap C) \equiv (A \cap B) \cap C$
    $A \cup (B \cup C) \equiv (A \cup B) \cup C$

5. $A \cap A = \emptyset$
    $A \cup A = A$
    $A - A = \emptyset$
    $A \oplus A = \emptyset$

3. $A \cap (B \cup C) \equiv (A \cap B) \cup (A \cap C)$
    $A \cup (B \cap C) \equiv (A \cup B) \cap (A \cup C)$

Proof. We may translate the identities using the language of logical operators, and the proof is a simple checking of truth tables. $\triangledown$

2.2.3   Power Sets and Direct Product

The concept of a set being part of another, bigger set is a useful set relation. We shall call the smaller set a subset of the bigger. More precisely,

Definition. A set A is a *subset* of a set B, denoted by $A \subseteq B$, when the following proposition holds for every element x.

$$x \in A \rightarrow x \in B$$

This definition simply says that $A \subseteq B$ exactly when the set A is contained in the set B, in the sense that every element of A also belongs to B. For example, $\mathbb{N} \subseteq \mathbb{Z} \subseteq \mathbb{Q} \subseteq \mathbb{R}$.

Question. How would you draw the Venn diagram in order to show the relation $A \subseteq B$?

Test 2.29. For arbitrary sets A and B, which proposition is false?

a) $A - B \subseteq A \cup B$
b) $A - B \subseteq A \oplus B$
c) $A \cap B \subseteq A \oplus B$
d) $A \oplus B \subseteq A \cup B$

Test 2.30. Suppose that $A \subseteq B$. Which one of the following sets is necessarily empty?

a) $A \cup B$
b) $A \cap B$
c) $A - B$
d) $A \oplus B$

Exercise 2.31. Suppose that $A \subseteq B$ and $B \subseteq C$. Explain logically why we then have $A \subseteq C$.

Note that by the definition of the operator if-then, both the relations $A \subseteq A$ and $\emptyset \subseteq A$ are always true, for any set A.

Question. Is it true that $\emptyset \subseteq \emptyset$?

Theorem 2.7. Let A and B be any two sets. Then $A = B$ if and only if $A \subseteq B$ and $B \subseteq A$.

Proof. Assume first that $A = B$. Then the only-if statement reduces to $A \subseteq A$, which is true. Conversely, assume that $A \subseteq B$ and $B \subseteq A$ both hold. By the definition of subset, if $x \in A$ then $x \in B$, and vice versa. It follows that $x \in A \leftrightarrow x \in B$, hence $A = B$. ▽

Example. Suppose that $A \cap B = \emptyset$. Use Theorem 2.7 to prove that $A \oplus B = A \cup B$.

Solution. It is clear that, in general, $A \oplus B \subseteq A \cup B$. Hence, it suffices to show that $A \cup B \subseteq A \oplus B$. Let $x \in A \cup B$. Then either $x \in A$ or $x \in B$, but not both since A and B are disjoint. Hence, $x \in A \oplus B$. This shows that $A \cup B \subseteq A \oplus B$, and consequently $A \oplus B = A \cup B$.

Exercise* 2.32. Suppose that $A \subseteq B$. Use Theorem 2.7 to prove that $A \oplus B = B - A$.

Definition. If A is a set, the *power set* of A, denoted by $P(A)$, is the set consisting of all the subsets of A, i.e., $P(A) = \{S \mid S \subseteq A\}$.

Note that this is the first time we introduce a set whose elements are again sets.

Example. Find $P(A)$ for the set $A = \{1, 2\}$.

Solution. We first identify all the subsets of A—there are four of them, i.e., $\{1\}$, $\{2\}$, and $A = \{1, 2\}$ and $\emptyset$. Hence, $P(\{1, 2\}) = \{\{1\}, \{2\}, \{1, 2\}, \emptyset\}$.

Exercise 2.33. Find $P(A)$.

a) $A = \{a, b, c\}$
b) $A = \{2, 3, 4, 5\}$
c) $A = \emptyset$
d) $A = \{x, \{7\}\}$

Question. Is it true that $P(\emptyset) = \emptyset$?

Having done the last exercise, you may have deduced the result of the next theorem, which explains why the name *power* set.

Theorem 2.8. If a set A consists of n elements, then A has exactly 2^n subsets, i.e., $P(A)$ is a set of 2^n elements.

Proof. Let $A = \{a_1, a_2, \ldots, a_n\}$. A subset $S \subseteq A$ can be represented by $\{s_1, s_2 \ldots, s_n\}$, where $s_i = 0$ if $a_i \notin S$ and $s_i = 1$ if $a_i \in S$. It is then clear that there are 2^n possible representations, hence that many subsets of A. $\triangledown$

Definition. The *cardinality* of a set A, denoted by $|A|$, is the number of elements in A, if finite.

Hence for example, $|\{a, b, c\}| = 3$ and $|\emptyset| = 0$. With this new notation we may rewrite Theorem 2.8 using the following proposition.

$$|A| = n \rightarrow |P(A)| = 2^n$$

Exercise 2.34. Evaluate $|P(A)|$.

a) $A = \{1, 2, 3, 4\} \cup \{3, 4, 5, 6\}$
b) $A = \{1, 2, 3, 4\} \oplus \{3, 4, 5, 6\}$
c) $A = P(\emptyset)$
d) $A = P(P(\emptyset))$

Question. What are the elements of the set $P(P(P(\emptyset)))$?

Definition. If A and B are two given sets, the *direct product* $A \times B$, read *A cross B*, is defined by $A \times B = \{(a, b) \mid a \in A \wedge b \in B\}$.

For example, if $A = \{1, 2\}$ and $B = \{3, 4\}$ then

$$A \times B = \{(1, 3), (1, 4), (2, 3), (2, 4)\}$$
$$B \times A = \{(3, 1), (3, 2), (4, 1), (4, 2)\}$$

Test 2.35. With $A = \{1, 2, 3, 4\}$ and $B = \{3, 4, 5\}$, what is $|P(A \times B)|$?

a) 32
b) 64
c) 128
d) 4096

Question. Is it true that $\emptyset \times \emptyset = \emptyset$?

Note that an element of the form (a, b) is treated as an ordered pair, e.g., $(1, 4) \neq (4, 1)$—just like what we have in the Cartesian coordinate system. Hence, in general, we have $A \times B \neq B \times A$.

Exercise* 2.36. Is it possible to have $A \times B = B \times A$ sometimes? Think of an example, other than $A = B$, or explain why it is not possible.

Exercise 2.37. For each claim below, investigate true or false.

a) $|A \times B| = |B \times A|$
b) $A \times (B \cup C) = (A \times B) \cup (A \times C)$
c) $P(A \times B) = P(A) \times P(B)$
d) $|P(A \times B)| = |P(A) \times P(B)|$

Theorem 2.9. If $|A| = m$ and $|B| = n$, then $|A \times B| = mn$.

Proof. An element $(a, b) \in A \times B$ can be selected from any one of the m elements in A, for a, and from n choices for b. Hence there are exactly mn such ordered pairs. ∇

2.3 Techniques of Proof

Proving a mathematical statement is an art of writing. There is no strict convention as to how a proof should look like. However, there are commonly accepted methods of proof which follow certain laws of logic. We look into a few of these methods, trying wherever possible to communicate in the language of propositions as we have learned it in this chapter.

2.3.1   Direct Proof and Contrapositive

Recall the definitions of even and odd numbers given at the beginning of Chapter 1. We will use these numbers to illustrate our first proof technique, called *direct proof*. The technique of direct proof applies to statements in the form of an implication $p \to q$. It begins by assuming p and shows, through a succession of implications, that q inevitably follows.

Example. Prove that if x is even then x^2 is also even.

Solution. If we let p denote the statement "x is even" and q the statement "x^2 is even," then we are to prove the proposition $p \to q$.

$\quad p : x$ is even

$\quad\quad \to x = 2n$ for some $n \in \mathbb{Z}$ $\qquad\qquad$ (definition of even numbers)

$\quad\quad \to x^2 = 4n^2$ $\qquad\qquad\qquad\qquad$ (by squaring both sides)

$\quad\quad \to x^2 = 2(2n^2)$

$\quad\quad \to x^2 = 2m$ where $m = 2n^2 \in \mathbb{Z}$ $\qquad$ (since n is integer)

$\quad\quad \to x^2$ is even $: q$ $\qquad\qquad\qquad\qquad$ (by definition)

Example. Prove that the product of two odd numbers is again odd.

Solution. This statement does not look like an implication, but it really is. Simply let $p : x$ and y are odd and $q : xy$ is odd. The proposition to be proved is essentially $p \to q$.

$\quad\quad p : x$ and y are odd

$\quad\quad\quad \to x = 2n+1$ and $y = 2m+1$ with both $m, n \in \mathbb{Z}$

$\quad\quad\quad \to xy = (2n+1)(2m+1)$

$\quad\quad\quad \to xy = 4nm + 2n + 2m + 1$

$\quad\quad\quad \to xy = 2(2nm + n + m) + 1$

$\quad\quad\quad \to xy = 2k+1$ where $k = 2nm + n + m \in \mathbb{Z}$

$\quad\quad\quad \to xy$ is odd $: q$

Question. Is it wrong if we let $x = 2n+1$ and $y = 2n+1$ in the above?

Exercise 2.38. Prove the following statements using direct proof.

a) If x is odd, then x^3 is also odd.
b) If x is even, then so is $x^2 - 4x + 2$.
c) The sum of two odd numbers is even.
d) The sum of two rational numbers is again rational.

There are times when direct proof may not be the easiest way to establish $p \to q$. In such cases, the contrapositive of this implication is often a useful substitute for the statement before we attempt to prove it. If you recall, Theorem 2.2 states that

$$p \to q \equiv \neg q \to \neg p$$

Example. Let x be an integer. Prove that if x^2 is even then so is x.

Solution. As before, we let $p : x^2$ is even and $q : x$ is even. We wish to prove the validity of $p \to q$. Direct proof would start with $x^2 = 2n$ and have us show that $x = \sqrt{2n}$ is twice an integer. That would be hard. To circumvent this difficulty, we shall instead prove the equivalent statement $\neg q \to \neg p$.

$$\neg q : x \text{ is odd} \quad \text{(the integer } x \text{ is } not \text{ even)}$$
$$\to x = 2n + 1 \text{ for some } n \in \mathbb{Z}$$
$$\to x^2 = 4n^2 + 4n + 1$$
$$\to x^2 = 2(2n^2 + 2n) + 1$$
$$\to x^2 = 2m + 1 \text{ where } m = 2n^2 + 2n \in \mathbb{Z}$$
$$\to x^2 \text{ is odd} : \neg p$$

At this point it is appropriate to remark that writing a mathematical proof need not be so formal as to represent every statement using p and q. The next example is simply meant to show how our writing style can be more casual for the sake of better readability.

Example. Prove that if $2n$ is an irrational number, then n is too.

Solution. We use contrapositive. Suppose that n is rational. We may write $n = a/b$ for some integers a, b. Then $2n = 2a/b$, which shows that $2n$ is also a rational number. This completes the proof.

Exercise 2.39. Prove the following statements using contrapositive.
a) If x^3 is odd, then x is also odd.
b) If $x^2 - 3$ is irrational, so is $x - 3$.
c) If the sum of two integers is odd, then one of them is odd.
d) If the product of two integers is even, then one of them is even.

Question. How can one solve Exercise 2.39(d) using direct proof, instead of contrapositive, plus the fact that 2 is prime?

Exercise 2.40. If x is odd, then $x^2 - 1$ is divisible by 8. Elias proved this statement as follows. If x is even, then $x = 2n$. And therefore $x^2 - 1 = 4n^2 - 1$ is an odd number, not divisible by 8. Please explain to Elias why his proof is fundamentally wrong, then write the correct proof.

Exercise* 2.41. Explain why, if $p > 2$ is a prime number, then p is odd. Then use this knowledge to prove that if $p \bmod 3 = 1$, then $p \bmod 6 = 1$.

2.3.2   Proof by Cases

Consider the following proof problem and its solution.

Example. Prove that the quantity $x^2 - 3x + 2$ is even for any $x \in \mathbb{Z}$.

Solution. Let $p : x \in \mathbb{Z}$ and $q : x^2 - 3x + 2$ is even. We consider two cases: x is even and x is odd.

1. Let $p_1 : x$ is even. We shall establish $p_1 \to q$ using direct proof. Let $x = 2n$ with $n\mathbb{Z}$. Then $x^2 - 3x + 2 = 4n^2 - 6n + 2 = 2m$, where $m = 2n^2 - 3n + 1 \in \mathbb{Z}$. Hence, $x^2 - 3x + 2$ is even.

2. Let $p_2 : x$ is odd. We will also prove $p_2 \to q$. Let $x = 2n + 1$ for some $n \in \mathbb{Z}$. Then $x^2 - 3x + 2 = 4n^2 - 2n = 2m$, where $m = 2n^2 - n \in \mathbb{Z}$. It follows that $x^2 - 3x + 2$ is even.

Since every integer is either even or odd, we have actually proved that $p \to q$.

This example illustrates the method of *proof by cases*. In this instance we have $p \equiv p_1 \lor p_2$, which enables us to substitute $p \to q$ by the proposition $(p_1 \to q) \land (p_2 \to q)$. Logically speaking, this is justified by the equivalence

$$(p_1 \lor p_2) \to q \equiv (p_1 \to q) \land (p_2 \to q)$$

Of course, there are times when p more conveniently breaks down to three or more cases, instead of just two, like in the next example.

Example. Prove that every odd number x can be written either in the form $x = 4n + 1$ or $x = 4n + 3$.

Solution. By Theorem 1.1, we have $0 \le x \bmod 4 \le 3$. This gives four possible cases, i.e., $x = 4n$, $x = 4n + 1$, $x = 4n + 2$, and $x = 4n + 3$, for some $n \in \mathbb{Z}$. The first case and the third case apply to an even number x. Hence if x is odd, then either $x = 4n + 1$ or $x = 4n + 3$.

Exercise 2.42. Use proof by cases to establish each claim below.
a) The number $x^2 + x$ is even for all integers x.
b) For any $x \in \mathbb{Z}$, the number $x^2 + 2$ is not a multiple of 4.
c) If $x \in \mathbb{Z}$ then $x^3 - x$ is a multiple of 3. (Hint: consider $x \bmod 3 = 0, 1, 2$.)
d) For any two real numbers, we have $|xy| = |x| \, |y|$. (Hint: consider the four cases where $x < 0$ or $x \ge 0$, and for y similarly.)

Exercise* 2.43. Prove that every prime number $p > 3$ can be written in the form $p = 6n \pm 1$.

2.3.3 Proving Equivalence

Proving an equivalence means proving a biconditional statement $p \leftrightarrow q$. Since we have the established fact that

$$p \leftrightarrow q \equiv (p \rightarrow q) \wedge (q \rightarrow p)$$

then in order to assert $p \leftrightarrow q$, it suffices to prove both $p \rightarrow q$ and $q \rightarrow p$.

Example. Prove that an integer x is even if and only if x^2 is even.

Solution. We are to prove $p \leftrightarrow q$, where $p : x$ is even and $q : x^2$ is even. Both parts, proving $p \rightarrow q$ and $q \rightarrow p$, have been demonstrated as the examples of Section 2.3.1. In particular, establishing $q \rightarrow p$ in this case is best done via contrapositive.

As a second example, this time without explicitly stating the propositions in terms of p and q, we state a useful theorem concerning the congruence $a \equiv b \pmod{n}$, which means that $a \bmod n = b \bmod n$. (See Section 1.2.1.)

Theorem 2.10. Let $a, b \in \mathbb{Z}$ and $n \in \mathbb{N}$. Then $a \equiv b \pmod{n}$ if and only if n divides $a - b$.

Proof. We note that, by definition, $a \bmod n = a - \lfloor \frac{a}{n} \rfloor n$ and quite similarly, $b \bmod n = b - \lfloor \frac{b}{n} \rfloor n$. It follows that

$$a \bmod n - b \bmod n = a - b - \left(\lfloor \tfrac{a}{n} \rfloor - \lfloor \tfrac{b}{n} \rfloor \right) n \qquad (2.1)$$

Hence, if $a \bmod n = b \bmod n$ then $a - b = \left(\lfloor \frac{a}{n} \rfloor - \lfloor \frac{b}{n} \rfloor \right) n$, which is a multiple of n since the floor function is integer valued.

Conversely, assume that n divides $a - b$. By Theorem 1.2, (2.1) implies that n also divides $a \bmod n - b \bmod n$. But from Theorem 1.1, we know that the value of a remainder mod n is between 0 and $n - 1$, inclusive. Hence, $-n < a \bmod n - b \bmod n < n$, and so to be divisible by n, we must have $a \bmod n - b \bmod n = 0$. That gives $a \bmod n = b \bmod n$. $\triangledown$

Question. What exactly is meant by the word *conversely*?

Exercise 2.44. Prove the following biconditional statements.

a) The number $x^2 - 4x + 2$ is odd if and only if x is.
b) The product of two numbers is odd if and only if both of them are.
c) The product of two integers is divisible by 5 if and only if one of them is.
d) The sum of two integers is odd if and only if one of them is odd and the other even.

Another example, involving sets, is even more subtle in the only-if part of the proof. Note that by contrapositive, in order to establish $p \leftrightarrow q$, we may simply prove "if p is true then q is true" and "if p is false then q is false." Moreover, $p \leftrightarrow q$ may well be replaced by $q \leftrightarrow p$.

Example. Suppose that A and B are two arbitrary non-empty sets. Prove that $A \times B = B \times A$ if and only if $A = B$.

Solution. The case $A = B$ is trivial. Suppose now $A \neq B$. We may assume, *without loss of generality,* that there exists $x \in A - B$. (If this assumption turns false, then there is $x \in B - A$, and we simply exchange the roles of A and B.) This will give an element $(x, y) \in A \times B$ such that $(x, y) \notin B \times A$, implying that $A \times B \neq B \times A$.

Exercise 2.45. With sets, prove that $A \oplus B = B - A$ if and only if $A \subseteq B$.

Exercise* 2.46. Let both $m, n \in \mathbb{N}$. Show that $\gcd(m, n) = \text{lcm}(m, n)$ if and only if $m = n$.

2.3.4   Proof by Contradiction

There are times when proving the validity of a proposition p is difficult, whereas verifying the absurdity of $\neg p$ is not as hard. This is an acceptable alternative proof technique, called *proof by contradiction.* As a first example, we demonstrate that the number $\sqrt{2}$ is irrational, thus keeping our promise made at the beginning of Chapter 1.

Example. Prove that the number $\sqrt{2}$ is irrational.

Solution. We will show that the negated statement "The number $\sqrt{2}$ is rational" is absurd, so we may conclude that the statement "The number $\sqrt{2}$ is irrational" is true.

Suppose that $\sqrt{2} \in \mathbb{Q}$. Then we may write $\sqrt{2} = a/b$, where $a, b \in \mathbb{Z}$ with no common factor except 1. Then $2 = a^2/b^2$, that is, $2b^2 = a^2$. It follows that the number a^2 is even, hence a is even. (Remember?) We now write $a = 2c$ for some $c \in \mathbb{Z}$. Substituting, we get $2b^2 = 4c^2$, which simplifies to $b^2 = 2c^2$. By similar reasoning, we see that b is as well even. This is absurd (a contradiction) since now a and b have 2 as a common factor, contradicting our assertion earlier that they have no common factor.

Exercise 2.47. Prove the following claims by way of contradiction.

a) The number $\sqrt[3]{3}$ is irrational.
b) The number $\sqrt[n]{p}$ is irrational when p is prime and $n \geq 2$.
c) The number $\log_{10} 2$ is irrational.

d) The sum of an irrational number and a rational is irrational.

Question. Is the sum of two irrational numbers again irrational?

Proof by contrapositive is really a form of contradiction. Suppose that we are to prove the compound statement $p \to q$. We assume, as in direct proof, that p holds and try to show that q must follow. Instead, we demonstrate why $\neg q$ is absurd (proof by contradiction) by showing that $\neg q \to \neg p$ (proof by contrapositive). And since $\neg p$ and p cannot both be true, we arrive at the desired contradiction.

Exercise 2.48. Show that if a and b are odd, then $x^2 = a^2 + b^2$ has no integer solution.

As a second example, we present again the proof that there exist infinitely many prime numbers. Differing from the proof of Theorem 1.10, however, this time we closely follow Euclid's original proof.

Example. Prove that the are infinitely many prime numbers.

Solution. We assume, by contradiction, that only $p_1, p_2, \ldots, p_n$ are prime numbers. Obviously the number $N = p_1 p_2 \cdots p_n + 1$ is larger than any of these primes, so by assumption N is composite. By the fundamental theorem of arithmetic (Theorem 1.8), one of these primes divides N. That same prime will divide $N - p_1 p_2 \cdots p_n = 1$, according to Theorem 1.2. This is absurd because the number 1 does not have any prime divisor.

Exercise* 2.49. Prove the following claims, in the given order.
a) If $a \bmod 4 = 1$ and $b \bmod 4 = 1$ then $ab \bmod 4 = 1$.
b) If $p > 2$ is a prime number, then $p \bmod 4 = 1$ or 3.
c) If $n \bmod 4 = 3$, then n has a prime factor p such that $p \bmod 4 = 3$.
d) There are infinitely many primes p such that $p \bmod 4 = 3$.

Exercise* 2.50. Use contradiction to prove that a Carmichael number must have at least three prime factors. See Section 1.4.5 for definition.

2.4 Predicates and Quantifiers

In Mathematics we come across a statement like $x^2 = 4$, whose truth value shall be determined by the variable x. If we let $P(x)$ stand for the statement $x^2 = 4$, then $P(2)$ becomes a proposition whose value in this case is true. Similarly, $P(3.14)$ will be false. This proposition function $P(x)$ is an example of what we call a *predicate*.

Question. Where have we seen a predicate $P(x)$ earlier in this chapter?

2.4.1   There Is and For All

A predicate can also become a proposition when prefixed by a *quantifier*. There are two quantifiers, i.e.,

$$\exists \quad \text{read: } \textit{there is} \text{ or } \textit{there exists}$$
$$\forall \quad \text{read: } \textit{for all} \text{ or } \textit{for every}$$

Example. Let $P(x) : x^2 = 4$ and $Q(x) : x^2 > 0$. Determine true or false.

a) $\exists x \in \mathbb{R} : P(x)$
b) $\forall x \in \mathbb{Z} : P(x)$
c) $\exists x \in \mathbb{Z} : Q(x)$
d) $\forall x \in \mathbb{R} : Q(x)$

Solution. Note as follows how each statement is supposed to read.

a) $\exists x \in \mathbb{R} : P(x)$ represents the statement "There is a real number x such that $x^2 = 4$." This statement is *true* because there does exist such a number $x \in \mathbb{R}$, e.g., $x = 2$. (In fact there is another example, $x = -2$, but producing one example is enough.)
b) $\forall x \in \mathbb{Z} : P(x)$ stands for "For all integers x, we have $x^2 = 4$. This is *false*, for example consider $x = 1$, which gives "$1^2 = 4$." (*Sometimes* this predicate can be true, e.g., for $x = 2$, but not *always* true.)
c) $\exists x \in \mathbb{Z} : Q(x)$ is true; just let $x = 3$, for instance. (In fact, there are abundantly many examples, for as long as $x \neq 0$.)
d) $\forall x \in \mathbb{R} : Q(x)$ is false, for when $x = 0$ we get "$0^2 > 0$" (even though $x = 0$ is the *only* instance for which $Q(x)$ becomes false).

Exercise 2.51. For each predicate $P(x)$ given below, determine the truth values of $\exists x \in \mathbb{R} : P(x)$ and $\forall x \in \mathbb{R} : P(x)$.

a) $P(x) : x > 2x$
b) $P(x) : x^4 < -1$
c) $P(x) : 3x^2 - 15x + 7c = 0$
d) $P(x) : 2x^2 - 9x + 11 > 0$

Exercise* 2.52. Let S be any set whose elements are again sets. Define the *generalized union* and *generalized intersection* of sets over S, respectively,

$$\bigcup_{A \in S} A = \{x \mid \exists A \in S : x \in A\} \quad \text{and} \quad \bigcap_{A \in S} A = \{x \mid \forall A \in S : x \in A\}$$

In particular, if $A_n = \{x \in \mathbb{R} \mid 0 \leq x \leq \frac{1}{n}\}$ then prove that

$$\bigcup_{n \in \mathbb{N}} A_n = A_1 \quad \text{and} \quad \bigcap_{n \in \mathbb{N}} A_n = \{0\}$$

Appendix: Reading Double Quantifiers

A predicate, like any other function, may well involve two or more variables. The predicate $P(x, y) : (x + y)^2 = x^2 + y^2$ is an example.

Question. For which values of $(x, y) \in \mathbb{R} \times \mathbb{R}$ is $P(x, y) : (x + y)^2 = x^2 + y^2$ true?

Example. Let $P(x, y) : x^2 \geq y^2$. Determine true or false, where $x, y \in \mathbb{R}$.

a) $\exists x \exists y : P(x, y)$
b) $\forall x \forall y : P(x, y)$
c) $\exists x \forall y : P(x, y)$
d) $\forall x \exists y : P(x, y)$
e) $\exists y \forall x : P(x, y)$

Solution. It is important to note how each statement reads.

a) $\exists x \exists y : P(x, y)$ represents the statement "There is a real number x and another number y such that $x^2 \geq y^2$. This is *true*, e.g., $x = 5$ and $y = 3$.
b) $\forall x \forall y : P(x, y)$ stands for "For all numbers x and all numbers y, we have $x^2 \geq y^2$. This is *false*, for example if $x = 3$ and $y = 5$.
c) $\exists x \forall y : P(x, y)$ reads "There is a number x such that $x^2 \geq y^2$ for all numbers y. The fact is there is *no* number x satisfying the condition, for say if $x = A$ then the statement $A^2 \geq y^2$ is not true when $y = |A| + 1$ and definitely not for all y. Hence this proposition is *false*.
d) $\forall x \exists y : P(x, y)$ reads "For every number x, there is a number y such that $x^2 \geq y^2$. This is actually *true* because given a number $x = A$, we can simply let $y = A$ to satisfy the inequality $A^2 \geq y^2$.
e) $\exists y \forall x : P(x, y)$ is a *true* statement, e.g., let $y = 0$ so that for all numbers x the inequality $x^2 \geq 0$ holds. (There are no other examples really, for if $y \neq 0$ then the proposition $\forall x : P(x, y)$ is false. But, we have agreed that one example suffices.)

Exercise 2.53. Let $P(x, y) : x^2 + y^2 > 5$. Determine true or false, assuming that all numbers are real numbers.

a) $\exists x \forall y : P(x, y)$
b) $\forall x \exists y : P(x, y)$
c) $\exists y \forall x : P(x, y)$
d) $\forall y \exists x : P(x, y)$

Exercise 2.54. Repeat the previous exercise using each predicate below.

a) $P(x, y) : x^2 - y^2 > 5$
b) $P(x, y) : x^2 - y > 5$
c) $P(x, y) : x^2 + y^2 = (x + y)^2$
d) $P(x, y) : x^2 < x + y^2$

Test 2.55. Which predicate makes the proposition $\exists y \forall x : P(x, y)$ true?

a) $P(x, y) : x^2 - y^2 > 0$
b) $P(x, y) : x^2 - y > 0$
c) $P(x, y) : x - y^2 > 0$
d) $P(x, y) : x - y > 0$

2.4.2   Proving Existential Statements

We see that, in order to verify a proposition of the form $\exists x : P(x)$, it suffices to find a particular value of x which will make the predicate true. To prove that $\exists x : P(x)$ is false, on the other hand, we have to demonstrate that the proposition $\forall x : \neg P(x)$ holds. Conversely, to show that $\forall x : P(x)$ is false, we essentially seek to establish $\exists x : \neg P(x)$. Intuitively, we have the following pair of logical equivalence.

$$\neg \exists x : P(x) \equiv \forall x : \neg P(x)$$
$$\neg \forall x : P(x) \equiv \exists x : \neg P(x)$$

Question. How do we rewrite the statement "Not all prime numbers are odd" using the quantifier *for all*?

Example. Prove that there is a prime number with unit digit 1, but never with 0. Moreover, prove that not all prime numbers are odd.

Solution. There is a prime with unit digit 1, e.g., $p = 11$. But if p has unit digit 0, then p is a multiple of 10, hence composite. Therefore, no prime has unit digit 0. To show that not all primes are odd, it suffices to find an even prime number, i.e., $p = 2$.

Exercise 2.56. Prove the following propositions.

a) There is a prime number p such that $p \bmod 4 = 1$ and $p \bmod 5 = 4$.
b) Not all quadratic equations $ax^2 + bx + c = 0$ have a root $x \in \mathbb{R}$.
c) There is no prime number p for which $\sqrt{p} \in \mathbb{N}$.
d) For all $x \in \mathbb{R}$, we have $x^2 - 6x + 11 > 0$.

Exercise* 2.57. Prove that the number $\log_2 3$ is irrational, and use this fact to show that there exist irrational numbers a and b such that a^b is rational, with the particular choice of $a = \sqrt{2}$.

At times we need to prove the *uniqueness* of the existence $\exists x : P(x)$, i.e., that $P(x)$ holds for one and only one element x. For this, the proposition $\exists ! x : P(x)$ reads "There exists a unique element x for which $P(x)$ holds. To prove $\exists ! x : P(x)$, we establish $\exists x : P(x)$ as well as $P(a) \wedge P(b) \to a = b$.

Example. Prove that there exists a unique set of cardinality zero. (This justifies the definition of *the* empty set $\emptyset$.)

Solution. Since $\forall x \in \mathbb{R} : x^2 \geq 0$, we see that the set $\{x \in \mathbb{R} \mid x^2 = -1\}$ has cardinality zero. This shows existence. To prove uniqueness, assume that A and B are two sets such that $|A| = |B| = 0$. Since the proposition $\exists x : x \in A$ has value false, then $A \subseteq B$ by the definition of a subset. Similarly, $B \subseteq A$. Hence, $A = B$ by Theorem 2.7.

Exercise 2.58. Prove that there is a unique $x \in \mathbb{R}$ such that $x^3 = -1$.

Exercise* 2.59. Prove that every positive rational number can be written as m/n, using a unique pair of natural numbers with $\gcd(m, n) = 1$.

2.4.3 Mathematical Induction

The technique of *mathematical induction* applies to a statement involving a predicate and the quantifier $\forall$ with the domain of positive integers. For example, consider the statement

$$1 + 2 + 3 + \cdots + n = \frac{n(n+1)}{2} \quad \text{for all } n \in \mathbb{N}$$

Here, the predicate $P(n) : 1 + 2 + 3 + \cdots + n = \frac{n(n+1)}{2}$ claims to hold for all integer values of $n \geq 1$. How do we prove such a statement? We need only establish the following two propositions.

    1) $P(1)$

    2) $P(n) \rightarrow P(n+1)$

Intuitively, the second statement, with $n = 1$ says that if $P(1)$ holds, so does $P(2)$. Since $P(1)$ holds, i.e., by (1), then $P(2)$ is true. But by (2) again, since $P(2)$ is true, so is $P(3)$. And again, $P(3)$ implies $P(4)$, then $P(5)$, and on for all $P(n)$, where $n \in \mathbb{N}$. Proving (2) is what we call the *induction step*.

Example. Prove that the identity $1 + 2 + 3 + \cdots + n = \frac{n(n+1)}{2}$ holds for all integers $n \geq 1$.

Solution. We let $P(n)$ denote this predicate,

$$P(n) : 1 + 2 + 3 + \cdots + n = \frac{n(n+1)}{2}$$

The statement we are to prove can be represented by $\forall n \geq 1 : P(n)$. Note that $P(5)$, for instance, stands for the proposition $1 + 2 + 3 + 4 + 5 = \frac{5(5+1)}{2}$, whose value is true. This is just an example. We proceed with the two parts of the proof.

1) $P(1)$ is the proposition $1 = \frac{1(1+1)}{2}$. Obviously then, $P(1)$ is true.

2) Using direct proof, we will show $P(n) \to P(n+1)$. Note first the statement $P(n+1) : 1 + 2 + 3 + \cdots + (n+1) = \frac{(n+1)(n+2)}{2}$, as we proceed.

$$P(n) : 1 + 2 + 3 + \cdots + n = \frac{n(n+1)}{2}$$

$$\to 1 + 2 + 3 + \cdots + n + (n+1) = \frac{n(n+1)}{2} + (n+1)$$

$$\to 1 + 2 + 3 + \cdots + n + (n+1) = \frac{n(n+1) + 2(n+1)}{2}$$

$$\to 1 + 2 + 3 + \cdots + n + (n+1) = \frac{n^2 + 3n + 2}{2}$$

$$\to 1 + 2 + 3 + \cdots + n + (n+1) = \frac{(n+1)(n+2)}{2}$$

$$\to P(n+1)$$

Exercise 2.60. Use induction to prove each identity below for all $n \in \mathbb{N}$.

a) $1 + 2 + 4 + \cdots + 2^{n-1} = 2^n - 1$
b) $1 + 3 + 9 + \cdots + 3^{n-1} = (3^n - 1)/2$
c) $1 + 4 + 9 + \cdots + n^2 = n(n+1)(2n+1)/6$
d) $1 + 9 + 25 + \cdots + (2n-1)^2 = n(2n-1)(2n+1)/3$

Question. How do we prove a proposition $\forall n \geq 2 : P(n)$ using induction?

Example. Prove that $2^n < n!$ whenever $n \geq 4$.

Solution. For $n = 4$, we have $2^4 < 4! = 24$, which is true. Now let $n \geq 4$. Assuming that $2^n < n!$, then $2^{n+1} = 2(2^n) < 2n! < (n+1)n! = (n+1)!$, which proves the inductive step.

Exercise 2.61. Use mathematical induction, where n is understood integer.

a) Prove that $3^n < n!$ whenever $n \geq 7$.
b) Prove that $3^n > 1 + 2^n$ provided that $n \geq 2$.
c) Prove that $n^2 < 2^n$ for every $n \geq 5$.
d) Prove that $n^n > n!$ for all $n \geq 2$.

Test 2.62. We wish to prove using induction that $2^n > n^3$ for all integers $n \geq k$. What value of k is most suitable?

a) 0
b) 1
c) 10
d) The statement is false.

Exercise 2.63. Prove for all $n \in \mathbb{N}$, using induction.
a) The number $2^{2n} - 1$ is a multiple of 3.
b) The number $n^3 + 2n$ is divisible by 3.
c) The number 5 divides $n^5 - n$.
d) The number $\frac{1}{7}(2^{n+2} + 3^{2n+1})$ is an integer.

For our final example on the proof by mathematical induction, we shall reestablish Theorem 2.8.

Example. Prove the proposition $\forall n \geq 0 : |A| = n \rightarrow |P(A)| = 2^n$.

Solution. For $n = 0$ of course, $A = \emptyset$, and A has exactly $2^0 = 1$ subset, namely A itself. We proceed by induction.

Assume that the theorem holds and consider a set A with $n+1$ elements, one of which we call $x \in A$. Divide the subsets of A in two groups: those which contain x and those which do not. Those which do not, are exactly the subsets of $B = A - \{x\}$. Since B consists of n elements, B has 2^n subsets according to our hypothesis. Moreover, the other group has 2^n subsets too because they are given by $S \cup \{x\}$ for every subset $S \subseteq B$. Hence A has a total of $2^n + 2^n = 2^{n+1}$ subsets, proving the induction step.

Exercise* 2.64. Use induction to prove the following set identities involving generalized union and intersection, defined in Exercise 2.52.

$$\neg \bigcup_{n \in \mathbb{N}} A_n = \bigcap_{n \in \mathbb{N}} \neg A_n \quad \text{and} \quad \neg \bigcap_{n \in \mathbb{N}} A_n = \bigcup_{n \in \mathbb{N}} \neg A_n$$

Exercise* 2.65. Using mathematical induction, prove again the fact that every integer $n \geq 2$ can be written as a product of prime numbers. This statement is part of the fundamental theorem of arithmetic (Theorem 1.8). Although logically equivalent, note and explain a significant modification on the inductive step of the proof.

Books to Read

1. M. Aigner and G. M. Ziegler, *Proofs from THE BOOK*, Fourth Edition, Springer 2010.

2. J.M. De Koninck and A. Mercier, *1001 Problems in Classical Number Theory*, American Mathematical Society 2007.

3. G. Pólya, *How to Solve It: A New Aspect of Mathematical Method*, Second Edition, Princeton University Press 1988.

4. D. Solow, *How to Read and Do Proofs: An Introduction to Mathematical Thought Processes*, Fifth Edition, Wiley 2009.

Chapter 3

Topics in Set Theory

The language of set theory plays a fundamental role in much of modern mathematics. We will study the idea of relations between elements of two sets. In particular, the concept of a function from a set to another will enable us to define the cardinality of infinite sets. We then close with an introduction to group theory, which is the beginning of the modern study of abstract algebra.

3.1 Binary Relations

Definition. Let A and B be two sets. A *relation* R from A to B means a subset $R \subseteq A \times B$.

For example, the following are some, but not all, binary relations from the set $\{0, 1\}$ to the set $\{x, y, z\}$.

a) $\{(0, x), (1, y), (0, z)\}$
b) $\{(0, x), (0, z), (1, x), (1, z)\}$
c) $\{(1, x), (1, y), (1, z)\}$
d) $\{(0, x), (1, x), (0, y), (1, y), (0, z), (1, z)\}$

Test 3.1. How many different binary relations from $\{0, 1\}$ to $\{x, y, z\}$ can we have in all?

a) 8
b) 9
c) 32
d) 64

The adjective *binary* indicates that there are two sets involved. Since we are not interested in studying relations with more than two sets, from now on we agree that the term *relation* always refers to a binary relation.

Since relations are sets, with two relations R and S we are allowed to operate on them, e.g., using the operator union or intersection. We now introduce a new set operator which is customized to relations.

Definition. Suppose that $R \subseteq A \times B$ and $S \subseteq B \times C$ are two relations. Then $S \circ R$ is the relation from A to C given by

$$S \circ R = \{(a, c) \mid (a, b) \in R \land (b, c) \in S\}$$

The notation $S \circ R$ is read R *circle* S (yes, right to left!) and we refer to this set operation as the *composition* of R with S.

Example. Let $A = \{1, 2, 3, 4\}$, $B = \{x, y, z\}$, and $C = \{4, 5, 9\}$. Consider two relations R and S given below, and find the elements of the relation $S \circ R \subseteq A \times C$.

$$R = \{(1, y), (1, z), (2, x), (2, y), (4, z)\} \subseteq A \times B$$
$$S = \{(x, 4), (x, 9), (y, 5), (z, 5), (z, 9)\} \subseteq B \times C$$

Solution. The first element $(1, y) \in R$ *matches* with the element $(y, 5) \in S$, resulting in the new element $(1, 5) \in S \circ R$. Next, $(1, z) \in R$ and $(z, 5) \in S$ yield the same $(1, 5)$, whereas $(1, z)$ and $(z, 9)$ give $(1, 9)$. In all, seven elements are composed in this manner which make up the resulting set.

$$S \circ R = \{(1, 5), (1, 9), (2, 4), (2, 5), (2, 9), (4, 5), (4, 9)\}$$

Definition. By the *inverse* of a relation $R \subseteq A \times B$, we mean the relation from B to A given by $R^{-1} = \{(b, a) \mid (a, b) \in R\}$.

For example, the inverse of $R = \{(1, 0), (5, 5), (9, -2)\} \subseteq \mathbb{N} \times \mathbb{Z}$ is the relation $R^{-1} = \{(0, 1), (5, 5), (-2, 9)\} \subseteq \mathbb{Z} \times \mathbb{N}$.

Exercise* 3.2. Given two relations, $R \subseteq A \times B$ and $S \subseteq B \times C$, prove the set identity $(S \circ R)^{-1} = R^{-1} \circ S^{-1}$.

3.1.1 Relations on a Set

For the time being, we shall focus only on relations $R \subseteq A \times B$ in the special case where $A = B$.

Definition. Let A be a given set. A relation R *on* A means a relation from A to itself, i.e., $R \subseteq A \times A$. In this case, we define $R^2 = R \circ R$, $R^3 = R \circ R^2$, and by induction, $R^n = R \circ R^{n-1}$ and $R^{-n} = (R^{-1})^n$ for all $n \in \mathbb{N}$.

Question. If $|A| = n$, how many relations on A are possible in all?

Exercise 3.3. Let $R = \{(1,3),(2,2),(2,4),(3,1),(4,3)\}$ be a relation on the set $A = \{1,2,3,4\}$. Find the elements of each relation given by the following compositions.
a) R^2
b) R^4
c) R^{-2}
d) $R \circ R^{-1}$

Question. Is it true that $R^4 = R^2 \circ R^2$?

Exercise* 3.4. Prove that $R^n \circ R^m = R^{m+n}$ for all $m, n \in \mathbb{N}$.

Definition. Given a set A, we define the *identity relation* on A to be the special relation $A^0 = \{(a,a) \mid a \in A\}$. Any relation R on A is then called

1) *reflexive* if $A^0 \subseteq R$.

2) *symmetric* if $R^{-1} = R$.

3) *anti-symmetric* if $R \cap R^{-1} \subseteq A^0$.

4) *transitive* if $R^2 \subseteq R$.

Question. What is the difference between A^0 and $A \times A$?

Exercise* 3.5. For any $R \subseteq A \times A$, show that $R \circ A^0 = R = A^0 \circ R$.

Example. Let $A = \{1,2,3,4\}$. For each relation $R \subseteq A \times A$ given below, determine whether R is reflexive, symmetric, anti-symmetric, or transitive.
a) $R = \{(1,1),(1,2),(2,1),(2,2),(2,4),(3,3),(4,2)\}$
b) $R = \{(1,1),(1,3),(2,2),(2,4),(3,1),(3,3),(4,2),(4,4)\}$
c) $R = \{(a,b) \in A \times A \mid a \leq b\}$
d) $R = \{(a,b) \in A \times A \mid a \bmod b = 1\}$

Solution. We note that $A^0 = \{(1,1),(2,2),(3,3),(4,4)\}$.

1) R is symmetric since $R^{-1} = R$ but not reflexive as $(4,4) \notin R$. Anti-symmetric is false, e.g., $(1,2) \in R \cap R^{-1}$. So is transitive false, because the composition of $(4,2)$ with $(2,4)$ yields $(4,4) \notin R$.

2) You can check that R is reflexive, symmetric, and transitive. Only anti-symmetric is false.

3) R is reflexive since $a \leq a$ for all $a \in A$. Now if $a \neq b$, either $a < b$ or $b < a$ but never both. It follows that R is anti-symmetric, but not symmetric. Lastly, R is transitive for if $a \leq b$ and $b \leq c$, then $a \leq c$.

4) $R = \{(1,2),(1,3),(1,4),(3,2),(4,3)\}$. This relation is anti-symmetric, but is neither symmetric nor reflexive. R is not transitive. (Why?)

Instead of using set notation, we may use the language of propositional logic to redefine the above definitions; we state the results as a theorem but leave the proof as a straightforward exercise.

Theorem 3.1. Let R be a relation on a set A. Then

1) R is reflexive if and only if $\forall a \in A : (a,a) \in R$.

2) R is symmetric if and only if $\forall a,b \in A : (a,b) \in R \to (b,a) \in R$.

3) R is anti-symmetric if and only if $(a,b) \in R \to (b,a) \notin R$ for all $a,b \in A$ with $a \neq b$.

4) R is transitive if and only if $(a,b) \in R \wedge (b,c) \in R \to (a,c) \in R$ for all $a,b,c \in A$.

Exercise 3.6. Determine whether each relation R on A is reflexive, symmetric, anti-symmetric, or transitive.
a) $A = \{2,4,6,8\}$ and $R = \{(a,b) \in A \times A \mid a+b > 4\}$
b) $A = \{2,4,6,8\}$ and $R = \{(a,b) \in A \times A \mid a \bmod b = 0\}$
c) $A = \{2,4,7,8,9\}$ and $R = \{(a,b) \in A \times A \mid a+b \text{ is even}\}$
d) $R = \{(a,b) \in \mathbb{Z} \times \mathbb{Z} \mid a < b\}$

Question. Is anti-symmetric the negation of symmetric?

Exercise* 3.7. Is it possible to have $R \subseteq A \times A$ which is both symmetric and anti-symmetric? Think of an example or explain why it is not possible.

Exercise 3.8. Let $A = \{1,2,3,4\}$. Give an example of a relation on A which satisfies the following properties.
a) reflexive (T) symmetric (F) anti-symmetric (T) transitive (F)
b) reflexive (F) symmetric (F) anti-symmetric (F) transitive (F)
c) reflexive (F) symmetric (T) anti-symmetric (F) transitive (T)
d) reflexive (T) symmetric (T) anti-symmetric (F) transitive (T)

3.1.2 Digraphs and Zero-One Matrices

A relation R on A can be visually represented by a graph which is called the digraph of R, defined as follows.

Definition. Let R be a relation on a set A. The *digraph* of R is a picture in which the elements of A are drawn as dots (points) and each element $(a,b) \in R$ is represented by a line connecting them, with direction from a to b. We call the dots *vertices* (sometimes, *nodes*) and the lines *edges*. An edge $(a,a) \in R$ is also called a *loop*.

Example. Let $A = \{1, 2, 3, 4\}$ and $R = \{(1, 3), (2, 1), (2, 4), (3, 3), (4, 2)\}$. In drawing the digraph, we spread out the four vertices somewhat evenly, for the mere sake of better visibility. The five edges, one of which is a loop, are put in place accordingly, but note that the edges $(2, 4)$ and $(4, 2)$ are displayed as a single line with two directions.

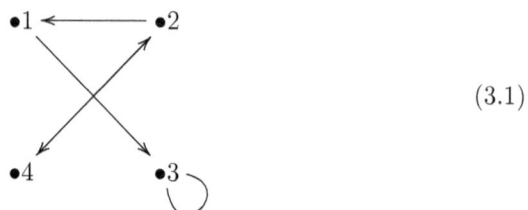

$$(3.1)$$

Question. Looking at the digraph, how do we know if a relation is reflexive, symmetric, anti-symmetric, or transitive?

Exercise 3.9. Redo Exercise 3.8, presenting your solutions in digraphs.

Another way to represent a relation $R \subseteq A \times A$ is by use of a matrix. Recall that a matrix is a two-dimensional array of elements, whose entry in the ith row and jth column will be denoted by m_{ij}, or simply (i, j). For example, a matrix M with 3 rows and 4 columns, otherwise called a 3×4 matrix, can be represented as follows.

$$M = \begin{bmatrix} (1,1) & (1,2) & (1,3) & (1,4) \\ (2,1) & (2,2) & (2,3) & (2,4) \\ (3,1) & (3,2) & (3,3) & (3,4) \end{bmatrix}$$

Definition. Suppose that the elements of A have been enumerated so that we may write $A = \{a_1, a_2, a_3, \ldots, a_n\}$. A relation R on A can then be represented by the $n \times n$ *zero-one matrix* $[R]$ whose elements $(i, j) = 1$ if $(a_i, a_j) \in R$, and $(i, j) = 0$ otherwise.

Example. Let $R = \{(1, 1), (1, 3), (2, 1), (2, 3), (2, 4), (3, 3), (4, 2), (4, 3)\}$ be a relation on the set $A = \{1, 2, 3, 4\}$. These eight elements of R correspond to the 1's in the following 4×4 zero-one matrix—for instance, $(1, 3) \in R$ tells us that the first row, third column entry in $[R]$ is 1.

$$[R] = \begin{bmatrix} 1 & 0 & 1 & 0 \\ 1 & 0 & 1 & 1 \\ 0 & 0 & 1 & 0 \\ 0 & 1 & 1 & 0 \end{bmatrix}$$

Exercise 3.10. Find $[R]$ for the relation R given by the digraph in (3.1).

Question. From the zero-one matrix, how do we know if a relation is reflexive, symmetric, anti-symmetric, or transitive?

Test 3.11. Which relation is anti-symmetric and transitive?

a) $\begin{bmatrix} 1 & 0 & 1 & 0 \\ 1 & 1 & 1 & 1 \\ 0 & 0 & 1 & 0 \\ 0 & 1 & 1 & 1 \end{bmatrix}$
b) $\begin{bmatrix} 0 & 0 & 1 & 0 \\ 1 & 0 & 1 & 0 \\ 0 & 0 & 0 & 0 \\ 0 & 1 & 1 & 0 \end{bmatrix}$
c) $\begin{bmatrix} 0 & 1 & 1 & 1 \\ 1 & 0 & 1 & 1 \\ 1 & 1 & 0 & 1 \\ 1 & 1 & 1 & 0 \end{bmatrix}$
d) $\begin{bmatrix} 1 & 0 & 1 & 0 \\ 1 & 0 & 1 & 0 \\ 0 & 0 & 1 & 0 \\ 1 & 1 & 1 & 0 \end{bmatrix}$

Exercise* 3.12. Let $[R] = [(i,j)]$ denote the matrix of a relation R on a set $A = \{a_1, a_2, \ldots, a_n\}$. Find a formula to compute the entry $(i,j)_2$ belonging to $[R^2] = [(i,j)_2]$, the zero-one matrix of the relation R^2.

3.1.3 Transitive Closures

Elias wants to fly out of Amman to Santiago, Chile. He asks his travel agent to book him his airplane tickets. The first thing that the travel agent checks is whether there are direct flights. If there is none, perhaps Elias can make a transit somewhere to connect with another flight. Sometimes two transits or more may be necessary for a traveler to reach his desired destination.

If A denotes the set of all airports in the world, the element $(a,b) \in R \subseteq A \times A$ tells us that there is a direct flight from a to b. Traveling from Amman to Santiago is at all possible, by direct flight or transits, if and only if the element (Amman, Santiago) belongs to the transitive closure of R, i.e.,

Definition. Let R be a relation on some set A. By the *transitive closure* of R we mean the relation on A given by

$$\overline{R} = R \cup R^2 \cup R^3 \cup \cdots \cup R^n$$

where $n = |A|$.

If, for instance, $(a,b) \in R^3$ then flying from a to b can be done with (at worse) three connecting flights, i.e., two transit times. Hence $(a,b) \in \overline{R}$ if and only if it is possible to connect from a to b, possibly by involving a number of transits.

Question. Why don't we need R^{n+1} in the definition of transitive closure?

Example. Let $A = \{1, 2, 3, 4\}$ and $R = \{(1,3), (2,1), (2,4), (3,2), (4,4)\}$. Find the transitive closure of R.

Solution. We use the definition $R^n = R \circ R^{n-1}$ to find

$$R^2 = \{(1,2), (2,3), (2,4), (3,1), (3,4), (4,4)\}$$
$$R^3 = \{(1,1), (1,4), (2,2), (2,4), (3,3), (3,4), (4,4)\}$$
$$R^4 = \{(1,3), (1,4), (2,1), (2,4), (3,2), (3,4), (4,4)\}$$

Hence $\overline{R} = \{(1,1), (1,2), (1,3), (1,4), (2,1), (2,2), (2,3), (2,4), (3,1), (3,2),$
$(3,3), (3,4), (4,4)\} = A \times A - \{(4,1), (4,2), (4,3)\}$.

Exercise 3.13. Find the transitive closure of each relation given below, on the set $A = \{1,2,3,4\}$.

a) $R = \{(1,2), (2,1), (2,3), (3,4)\}$
b) $R = \{(1,1), (2,1), (2,4), (3,2), (4,3)\}$
c) $R = \{(1,1), (1,4), (2,1), (2,2), (3,3), (4,4)\}$
d) $R = \{(1,4), (2,1), (2,4), (3,2), (4,3)\}$

Test 3.14. The zero-one matrix of some relation R is provided below. Which matrix represents the transitive closure of R?

$$[R] = \begin{bmatrix} 0 & 0 & 0 & 1 \\ 1 & 0 & 0 & 0 \\ 0 & 0 & 1 & 1 \\ 0 & 1 & 0 & 0 \end{bmatrix}$$

a) $\begin{bmatrix} 1 & 1 & 0 & 1 \\ 1 & 1 & 0 & 1 \\ 1 & 1 & 1 & 1 \\ 1 & 1 & 0 & 1 \end{bmatrix}$
b) $\begin{bmatrix} 0 & 1 & 1 & 1 \\ 1 & 0 & 1 & 1 \\ 1 & 1 & 1 & 1 \\ 1 & 1 & 1 & 0 \end{bmatrix}$
c) $\begin{bmatrix} 1 & 1 & 1 & 1 \\ 1 & 1 & 1 & 1 \\ 1 & 1 & 1 & 1 \\ 1 & 1 & 1 & 1 \end{bmatrix}$
d) $\begin{bmatrix} 1 & 1 & 1 & 1 \\ 1 & 1 & 1 & 1 \\ 0 & 0 & 1 & 1 \\ 1 & 1 & 1 & 1 \end{bmatrix}$

Question. Given $[R]$, the zero-one matrix of some relation R, can you write a computer program which computes $[\overline{R}]$, the matrix of $\overline{R}$?

Theorem 3.2. Let R be a relation on A. The transitive closure of R is the smallest transitive relation on A which contains R as a subset.

Proof. We first show that $\overline{R}$ is transitive. Let (a, b) and (b, c) be two elements in $\overline{R}$. This means that $(a, b) \in R^i$ and $(b, c) \in R^j$ for some exponents $i, j \leq n$. It follows that $(a, c) \in R^{i+j}$, and hence $(a, c) \in R^k \subseteq \overline{R}$ for some $k \leq n$.

Now let S be another transitive relation on A such that $R \subseteq S$. To complete the proof, we will show that $\overline{R} \subseteq S$ by proving that $R^k \subseteq S$ for all $k \leq n$. By induction, suppose that we have established $R^k \subseteq S$. Then $R^{k+1} = R \circ R^k \subseteq S \circ S \subseteq S$ since S is transitive. $\triangledown$

Exercise 3.15. Prove that $\overline{R} = R$ if and only if R is transitive.

Question. What can we say about $\overline{\overline{R}}$, i.e., the transitive closure of $\overline{R}$?

Exercise* 3.16. In a similar way, we may define the *reflexive closure* and *symmetric closure* of $R \subseteq A \times A$ to be the smallest reflexive, respectively symmetric, relation on A which contains R. Find a formula to find the reflexive closure of a given R, and similarly for symmetric closure.

3.1.4 Equivalence Relations

Definition. A relation R on a set A is an *equivalence relation* if R is reflexive, symmetric, and transitive. If R is an equivalence relation, for each $a \in A$ we define the *equivalence class* of a to be the set $[a] = \{x \in A \mid (a, x) \in R\}$.

Example. Let $A = \{1, 2, 3, 4, 5, 6\}$ and $R = \{(a, b) \in A \times A \mid a + b \text{ is even}\}$. Show that R is an equivalence relation and find all the equivalence classes of A under this relation.

Solution. We have $R = \{(1, 1), (1, 3), (1, 5), (2, 2), (2, 4), (2, 6), (3, 1), (3, 3), (3, 5), (4, 2), (4, 4), (4, 6), (5, 1), (5, 3), (5, 5), (6, 2), (6, 4), (6, 6)\}$ and note that R is reflexive, symmetric, and transitive. The equivalence classes are

$$[1] = \{1, 3, 5\} \qquad [2] = \{2, 4, 6\} \qquad [3] = \{1, 3, 5\}$$
$$[4] = \{2, 4, 6\} \qquad [5] = \{1, 3, 5\} \qquad [6] = \{2, 4, 6\}$$

Hence there are only two distinct classes, i.e., $\{1, 3, 5\}$ and $\{2, 4, 6\}$.

Question. Is it always true that $a \in [a]$ for every equivalence class of $a \in A$?

Exercise 3.17. Prove that each R is an equivalence relation and find the equivalence classes.
a) $R = \{(a, b) \in \mathbb{Z} \times \mathbb{Z} \mid a + b \text{ is even}\}$
b) $A = \{1, 2, 3, 4\}$ and $R = \{(a, b) \in A \times A \mid a = b\}$
c) $A = \{0, 5, 8, 9, 10, 11\}$ and $R = \{(a, b) \in A \times A \mid a \bmod 3 = b \bmod 3\}$
d) $A = \{1, 2, 3, 6, 7, 9, 11, 12\}$ and $R = \{(a, b) \in A \times A \mid a \equiv b \pmod 4\}$

Test 3.18. Which zero-one matrix represents an equivalence relation?

a) $\begin{bmatrix} 1 & 0 & 0 & 1 \\ 0 & 1 & 0 & 1 \\ 0 & 0 & 1 & 1 \\ 1 & 1 & 1 & 1 \end{bmatrix}$
b) $\begin{bmatrix} 1 & 0 & 0 & 1 \\ 0 & 1 & 1 & 0 \\ 0 & 1 & 1 & 0 \\ 1 & 0 & 0 & 1 \end{bmatrix}$
c) $\begin{bmatrix} 0 & 1 & 1 & 1 \\ 1 & 0 & 1 & 1 \\ 1 & 1 & 0 & 1 \\ 1 & 1 & 1 & 0 \end{bmatrix}$
d) $\begin{bmatrix} 1 & 0 & 0 & 0 \\ 1 & 1 & 0 & 0 \\ 1 & 1 & 1 & 0 \\ 1 & 1 & 1 & 1 \end{bmatrix}$

Theorem 3.3. Let R be an equivalence relation on a set A. For any $a, b \in A$, the following four propositions are equivalent one to another.

$$a \in [b] \quad \leftrightarrow \quad b \in [a] \quad \leftrightarrow \quad (a, b) \in R \quad \leftrightarrow \quad [a] = [b]$$

Moreover, if $(a, b) \notin R$ then $[a] \cap [b] = \emptyset$.

Proof. We have $(a, b) \in R$ if and only if $b \in [a]$. Since R is symmetric, $(a, b) \in R$ if and only if $(b, a) \in R$. This yields the equivalence among the first three. Moreover, since $a \in [a]$, then $[a] = [b]$ implies $a \in [b]$. Conversely, if $(a, b) \in R$ then $x \in [a]$ implies $(a, x) \in R$ and $(b, x) \in R$ by transitivity. Hence, $x \in [b]$ and $[a] \subseteq [b]$. By a symmetrical argument, then $[a] = [b]$.

To see the last claim, we show its contrapositive: let $x \in [a] \cap [b]$. Since $(a, x) \in R$ and $(b, x) \in R$, then $(a, b) \in R$, by symmetry and transitivity. $\triangledown$

Theorem 3.3 says that the set A is *partitioned* into equivalence classes—the word partitioned means divided into disjoint subsets. The next theorem states that the congruence relation $a \equiv b \pmod{n}$ given in Section 1.2.1 is an equivalence relation on $\mathbb{Z}$—an important one, in fact.

Theorem 3.4. Let $n \geq 2$ be a fixed integer. The congruence relation $R = \{(a, b) \in \mathbb{Z} \times \mathbb{Z} \mid a \bmod n = b \bmod n\}$ is an equivalence relation on $\mathbb{Z}$ with exactly n *congruence classes* given by

$$[a]_n = \{kn + a \mid k \in \mathbb{Z}\}$$

for every integer a belonging in the interval $0 \leq a \leq n - 1$.

Proof. It is clear that R is reflexive, symmetric, and transitive. That there are n classes given by $[0], [1], \ldots, [n-1]$ is a consequence of Theorem 1.1 and Theorem 3.3. Moreover, $a \bmod n = b \bmod n$ if and only if n divides $a - b$, by Theorem 2.10, which holds exactly when $b = kn + a$ for any $k \in \mathbb{Z}$. $\triangledown$

For example, with $n = 2$, there are two classes of integers, i.e., the set $[0]_2$ of even numbers and $[1]_2$ of odd numbers. Similarly for $n = 3$, the set $\mathbb{Z}$ is partitioned into three congruence classes:

$$[0]_3 = \{\ldots, -3, 0, 3, 6, 9, 12, 15, \ldots\}$$
$$[1]_3 = \{\ldots, -2, 1, 4, 7, 10, 13, 16, \ldots\}$$
$$[2]_3 = \{\ldots, -1, 2, 5, 8, 11, 14, 17, \ldots\}$$

Note that every integer belongs to exactly one class.

3.1.5   Partial Order Relations

Definition. A relation R on a set A is called a *partial order relation* when R is reflexive, anti-symmetric, and transitive.

Question. So what is the difference between an equivalence relation and a partial order relation?

Test 3.19. Which zero-one matrix represents a partial order relation?

a) $\begin{bmatrix} 1 & 1 & 0 \\ 0 & 1 & 1 \\ 0 & 0 & 1 \end{bmatrix}$
b) $\begin{bmatrix} 1 & 1 & 0 \\ 0 & 1 & 1 \\ 1 & 0 & 1 \end{bmatrix}$
c) $\begin{bmatrix} 1 & 1 & 1 \\ 1 & 1 & 0 \\ 1 & 0 & 1 \end{bmatrix}$
d) $\begin{bmatrix} 1 & 0 & 1 \\ 0 & 1 & 1 \\ 0 & 0 & 1 \end{bmatrix}$

A partial order relation R may be represented by its *Hasse diagram*, which can be obtained from the digraph of R in four simple steps:

1) Do not draw loops.

2) Whenever $(a, b) \in R$ and $(b, c) \in R$, do not draw (a, c).

3) Relocate the vertices of R such that each edge points upward.

4) Do not show the direction of each edge, i.e., remove the arrowheads.

Example. Let $A = \{1, 2, 3, 12, 18\}$ and $R = \{(a, b) \in A \times A \mid b \bmod a = 0\}$. Show that R is a partial order relation and draw its Hasse diagram.

Solution. The fact that R is reflexive, anti-symmetric, and transitive can be easily seen from the digraph, and we show below how it appears following each of the four steps leading to the Hasse diagram of R.

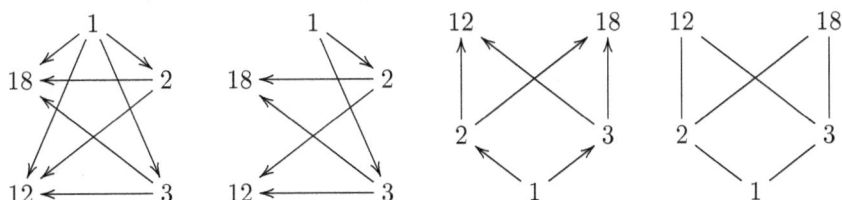

Exercise 3.20. Prove that each relation given below is a partial order relation and then draw its Hasse diagram.
a) $R = \{(a, b) \in \mathbb{N} \times \mathbb{N} \mid a \le b\}$
b) $A = \{2, 3, 12, 18, 36\}$ and $R = \{(a, b) \in A \times A \mid b \bmod a = 0\}$
c) $A = \{1, 2, 4, 8, 16\}$ and $R = \{(a, b) \in A \times A \mid b \bmod a = 0\}$
d) $A = \{1, 2, 3\}$ and $R = \{(X, Y) \in P(A) \times P(A) \mid X \subseteq Y\}$

Exercise* 3.21. Show that $R = \{(X, Y) \in P(A) \times P(A) \mid X \subseteq Y\}$ is a partial order relation on the power set of any set A.

Question. Suppose that the Hasse diagram has been found, as given below. How can we get back to the digraph or R?

(3.2)

Test 3.22. Which matrix corresponds to the Hasse diagram given in (3.2)?

a) $\begin{bmatrix} 1 & 0 & 0 & 0 \\ 1 & 1 & 0 & 0 \\ 0 & 1 & 1 & 1 \\ 1 & 0 & 0 & 1 \end{bmatrix}$
b) $\begin{bmatrix} 1 & 1 & 0 & 1 \\ 1 & 1 & 1 & 0 \\ 0 & 1 & 1 & 1 \\ 1 & 0 & 1 & 1 \end{bmatrix}$
c) $\begin{bmatrix} 1 & 0 & 0 & 0 \\ 1 & 1 & 0 & 0 \\ 1 & 1 & 1 & 1 \\ 1 & 0 & 0 & 1 \end{bmatrix}$
d) $\begin{bmatrix} 0 & 0 & 0 & 0 \\ 1 & 0 & 0 & 0 \\ 1 & 1 & 0 & 1 \\ 1 & 0 & 0 & 0 \end{bmatrix}$

Theorem 3.5. The relation $R = \{(a, b) \in \mathbb{N} \times \mathbb{N} \mid b \bmod a = 0\}$ is a partial order relation on $\mathbb{N}$.

Proof. It is clear that $(a, a) \in R$ for every $a \in \mathbb{N}$ because $a \bmod a = 0$. Hence R is reflexive. If $a \bmod b = 0 = b \bmod a$, then both b/a and $a/b \in \mathbb{N}$. It follows that $b/a = 1 = a/b$ and $a = b$, so R is anti-symmetric. Finally to show transitive, if (a, b) and $(b, c) \in R$, then b/a and $c/b \in \mathbb{N}$. Hence $b/a \times c/b = c/a \in \mathbb{N}$, implying that $(a, c) \in R$. ▽

Question. Can you partially sketch the Hasse diagram for this partial order relation $b \bmod a = 0$ on $\mathbb{N}$?

Definition. A relation R on a set A is called a *total order* if R is a partial order such that for any two elements $a, b \in A$, either $(a, b) \in R$ or $(b, a) \in R$.

Exercise 3.23. Which ones of the partial order relations given in Exercise 3.20 are total order relations?

Test 3.24. Which Hasse diagram below represents a total order relation?

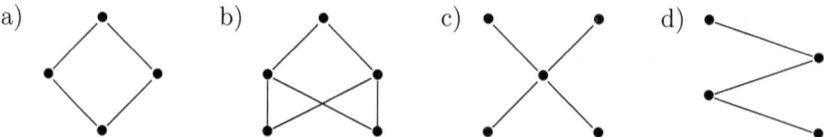

a)   b)   c)   d)

Question. Can you see why the Hasse diagram of a total order relation can always be drawn as a straight line?

Theorem 3.6. The relation $R = \{(a, b) \in \mathbb{R} \times \mathbb{R} \mid a \leq b\}$ is a total order relation on $\mathbb{R}$.

Proof. Reflexive is clear. To see why R is transitive, note that $a \leq b$ and $b \leq c$ imply $a \leq c$. Finally, if $a \neq b$ we have either $a < b$ or $b < a$ but never both, hence R is anti-symmetric and a total order relation. ▽

Exercise* 3.25. Suppose that R is a total order relation on a set A. Prove that the symmetric closure of R is $A \times A$. See Exercise 3.16 for definition.

3.2 Functions

We consider again the concept of a relation from a set A to another, possibly different, set B. The familiar notion of a function which is normally taught in Calculus can now be presented as a binary relation.

Definition. Let A and B be two given sets. The relation $f \subseteq A \times B$ is a *function* from A to B if there is a unique element $(a, b) \in f$ for every $a \in A$. In such a case, we write $f : A \rightarrow B$. The element $b \in B$ for which $(a, b) \in f$ will be denoted by $b = f(a)$. Thus $f = \{(a, f(a)) \mid a \in A\}$.

Question. How do we define a function using the symbols $\exists$ and $\forall$?

For example, the function $f : \mathbb{R} \rightarrow \mathbb{R}$ given by $f(x) = x^2$ refers to the relation $f = \{(x, x^2) \mid x \in \mathbb{R}\}$. The following are also some, but not all, possible functions from $\{1, 2, 3, 4\}$ to $\{a, b, c, d, e\}$.

a) $\{(1, a), (2, b), (3, c), (4, d)\}$
b) $\{(1, b), (2, c), (3, d), (4, e)\}$
c) $\{(1, b), (2, d), (3, b), (4, c)\}$
d) $\{(1, a), (2, a), (3, a), (4, a)\}$

Test 3.26. How many different functions from $\{1, 2, 3, 4\}$ to $\{a, b, c, d, e\}$ are possible?

a) 20
b) 120
c) 625
d) 1024

Question. By observing the zero-one matrix of a relation $R \subseteq A \times A$, how can we tell if R is a function?

Definition. The relation R on A given by $R = A^0$ defines a special function $\iota\delta_A : A \rightarrow A$ given by $\iota\delta_A(a) = a$ for every $a \in A$. Simply denoted by $\iota\delta$ if there is no ambiguity, this relation is called the *identity function* on A.

Definition. If $f : A \rightarrow B$ is a function, we call the set A the *domain* and B the *codomain* of f. Moreover, by the *range* of f we mean the subset of B given by $f(A) = \{f(a) \mid a \in A\}$.

For example, the function $f : \mathbb{R} \rightarrow \mathbb{R}$ given by $f(x) = x^2$ has $\mathbb{R}$ for both its domain and codomain. The range of this f is the non-negative interval $[0, \infty) = \{x \in \mathbb{R} \mid x \geq 0\}$.

Exercise 3.27. Give the largest possible domain for each function below, and then find the range as well.

a) $f(x) = \lfloor x \rfloor$
b) $f(x) = \lceil x \rceil - \lfloor x \rfloor$
c) $f(x) = \dfrac{\log(1 + x)}{\sqrt{x}}$
d) $f(m, n) = \gcd(m, n)$

Definition. A function $f : A \to B$ is *one-to-one* or *injective* if every $b \in B$ corresponds to at most one element $(a, b) \in f$. And f is *onto* or *surjective* when $f(A) = B$. If f is both injective and surjective, then we say that f is *bijective*. The noun *bijection* stands for a bijective function. Similarly, an *injection* (*surjection*) stands for an injective (surjective) function. The term *one-to-one correspondence* is a synonym for bijection.

Question. How do we define one-to-one using the symbols $\exists$ and $\forall$?

Note that a function $f : A \to B$ is injective if and only if the proposition $f(a) = f(a') \to a = a'$ holds for every two elements in A. For example, the function $f(x) = x^2$ is not one-to-one because, for instance, both $(2, 4) \in f$ and $(-2, 4) \in f$. The identity function $\iota\delta$, another example, is clearly a bijection on any domain A.

Exercise 3.28. Determine whether the function f is one-to-one or onto.
a) $f : \mathbb{R} \to \mathbb{R};\ f(x) = e^x$
b) $f : \mathbb{R} \to \mathbb{Z};\ f(x) = \lfloor x \rfloor$
c) $f : \mathbb{Z} \to \mathbb{Z};\ f(n) = n + 1$
d) $f : \mathbb{N} \times \mathbb{N} \to \mathbb{Z};\ f(m, n) = \gcd(m, n)$

Test 3.29. How many different one-to-one functions from $\{1, 2, 3, 4\}$ to $\{a, b, c, d, e, f\}$ are possible?
a) 24
b) 120
c) 360
d) 720

Test 3.30. How many different onto functions from $\{1, 2, 3, 4\}$ to $\{x, y, z\}$ are possible?
a) 6
b) 24
c) 33
d) 81

Question. If a function $f : A \to A$ is given by its zero-one matrix, how can we tell if it is one-to-one or onto?

Exercise* 3.31. Let $f : A \to B$ be a function. If $|A| = |B|$, and finite, prove that f is injective if and only if f is surjective.

Theorem 3.7. Consider two functions $f : A \rightarrow B$ and $g : B \rightarrow C$. The composition $g \circ f$ as a relation from A to C is again a function, i.e., the function $g \circ f : A \rightarrow C$ given by $g \circ f(a) = g(f(a))$ for every $a \in A$.

Proof. For every $a \in A$, there is a unique $(a, b) \in f$. With this $b \in B$, there is a unique $(b, c) \in g$. Hence every $a \in A$ corresponds to a unique $(a, c) \in g \circ f$, as desired. $\triangledown$

Exercise 3.32. Find $g \circ f$ by describing $g(f(x))$, where $x \in \mathbb{R}$ and $n \in \mathbb{Z}$, assuming some suitable domain and range for each one.

a) $f(x) = 2x - 1$; $g(x) = x^2 + 1$
b) $f(x) = 6 - 2x$; $g(x) = 3 - x/2$
c) $f(n) = 1/n$; $g(x) = 1/x$
d) $f(n) = n/(n+1)$; $g(x) = \lfloor x \rfloor$

Theorem 3.8. Let $f : A \rightarrow B$ and $g : B \rightarrow C$ be two functions. If both f and g are injective, or surjective, then $g \circ f$ is also injective, or surjective. Hence, if f and g are both bijective functions, so is $g \circ f$.

Proof. If f and g are onto, then $f(A) = B$ and $g(B) = C$. In that case, $g(f(A)) = C$ and $g \circ f : A \rightarrow C$ is onto. Suppose now $g(f(a)) = g(f(a'))$. If g is injective, then $f(a) = f(a')$. So if f is injective as well then $a = a'$, in which case $g \circ f$ is again an injection. $\triangledown$

Definition. Let $f : A \rightarrow B$ be a function. By the *inverse* of f we mean the inverse of f as a relation, i.e., $f^{-1} = \{(f(a), a) \mid a \in A\} \subseteq B \times A$. Note that f^{-1} may or may not be a function. If $S \subseteq B$, we define the *inverse image* of S under the function f to be the subset of A given by $f^{-1}(S) = \{a \in A \mid f(a) \in S\}$.

Exercise 3.33. Which ones of the functions f given in Exercise 3.28 have an inverse f^{-1} which is again a function?

Exercise 3.34. Find $f^{-1}(S)$ using the functions f given in Exercise 3.28, where S is given below for each one, respectively.

a) $S = (0, 1]$
b) $S = \{0\}$
c) $S = [0]_2$
d) $S = \{0\}$

Test 3.35. Which one of the following statements is generally false?

a) $f(S \cup T) = f(S) \cup f(T)$
b) $f(S \cap T) = f(S) \cap f(T)$
c) $f^{-1}(S \cup T) = f^{-1}(S) \cup f^{-1}(T)$

d) $f^{-1}(S \cap T) = f^{-1}(S) \cap f^{-1}(T)$

Theorem 3.9. Let $f : A \to B$ be a function. Then f^{-1} is again a function if and only if f is bijective, in which case $f^{-1} : B \to A$ is also a bijection. Moreover, in such a case, we have $f^{-1} \circ f = \iota\delta_A$ and $f \circ f^{-1} = \iota\delta_B$.

Proof. Suppose that f is bijective. Since f is onto, for every $b \in B$ there is $(a, b) \in f$ and is unique since f is one-to-one. Hence, $f^{-1} : B \to A$ is a function. Conversely, suppose that f^{-1} is a function. Then every $b \in B$ corresponds to a unique $a \in A$ such that $f(a) = b$. This forces f to be both onto and one-to-one.

Since $(f^{-1})^{-1} = f$, we see that if f^{-1} is a function, then the argument above shows that f^{-1} is bijective. Moreover, if $f(a) = b$ then $f^{-1}(f(a)) = f^{-1}(b) = a$, showing that $f^{-1} \circ f = \iota\delta_A$. Similarly, $f \circ f^{-1} = \iota\delta_B$.   ▽

3.3   Cardinal Numbers

Recall that the term cardinality is so far defined for finite sets. We will now extend this concept to include infinite sets as well. Before that, however, we note the following properties about the cardinality of finite sets.

Theorem 3.10. Suppose that both A and B are finite sets. Then

1) $|A| \leq |B|$ if and only if there is an injection from A to B.

2) $|A| \geq |B|$ if and only if there is a surjection from A to B.

3) $|A| = |B|$ if and only if there is a bijection from A to B.

4) if $|A| = |B|$ then any function from A to B is injective if and only if surjective.

Proof. Assume that $|A| = m$, $|B| = n$. Moreover let $A = \{a_0, a_1, \ldots, a_{m-1}\}$ and $B = \{b_0, b_1, \ldots, b_{n-1}\}$. If $m \leq n$, define $f : A \to B$ by $f(a_k) = b_k$. Clearly then f is injective, and bijective if $m = n$. If $m \geq n$, we define $f : A \to B$ by $f(a_k) = b_{k \bmod n}$, which is a surjection.

Now suppose that $f : A \to B$ is any function. If f is surjective then $|f^{-1}(B)| \geq n$, and since $f^{-1}(B) \subseteq A$, then $m \geq n$. If f is injective then $|f(A)| = m$, and since $f(A) \subseteq B$, then $m \leq n$. If furthermore f is bijective, then $f(A) = B$ and $m = n$.

Lastly, let $m = n$. If f is one-to-one then $|f(A)| = |B|$, so $f(A) = B$ and f is onto. But if f is not one-to-one then $|f(A)| < |B|$, so f is not onto. ▽

Based upon these observations, we redefine the term cardinality, now for arbitrary sets, as follows.

Definition. Every set S is associated with a *cardinal number,* which is called the *cardinality* of S and denoted by $|S|$, where we define the following relations involving two arbitrary sets A and B.

1) $|A| = |B|$ if there exists a bijection from A to B.

2) $|A| \preceq |B|$ if there exists an injection from A to B.

3) $|A| \prec |B|$ if $|A| \preceq |B|$ and $|A| \neq |B|$.

4) $|A| \succeq |B|$ if $|B| \preceq |A|$, and $|A| \succ |B|$ if $|B| \prec |A|$.

Intuitively the relation $|A| \preceq |B|$ says that B has "more" elements than A has, by means of one-to-one element matching. Since we are not actually counting the number of elements, this concept applies for infinite sets just as well as for finite sets.

While this definition of cardinality coincides with the old one for finite sets, justifying the use of the same name, the new notation $\preceq$ for "less than" is to keep us reminded that cardinal numbers are not assumed to obey the same law of ordering which applies to ordinary numbers.

Question. Why is it true that $|A| = |B|$ if and only if $|B| = |A|$?

Exercise 3.36. Prove that the relation $\preceq$ on cardinal numbers is reflexive, i.e., $|A| \preceq |A|$, and transitive, i.e., $|A| \preceq |B| \wedge |B| \preceq |C| \rightarrow |A| \preceq |C|$.

Definition. Denote the cardinality of $\mathbb{N}$ by $|\mathbb{N}| = \aleph_0$ (read *aleph naught*) and call a set A *countable* if $|A| \preceq \aleph_0$.

For example, the set $\mathbb{N}$ itself is countable, as is any subset $S \subseteq \mathbb{N}$ because the identity function $\iota\delta : S \rightarrow \mathbb{N}$ is certainly one-to-one. In fact, if $A \subseteq B$ then clearly $|A| \preceq |B|$.

Note that if $f : A \rightarrow \mathbb{N}$ is one-to-one, then the elements of $f(A)$ can be enumerated from least to greatest, $f(a_1) < f(a_2) < f(a_3) < \cdots$ Therefore, if A is countable, we may write $A = \{a_1, a_2, a_3, \ldots\}$ both for the finite as well as infinite cases.

Test 3.37. Which one of the following sets is countable?

a) $\{x \in \mathbb{Z} \mid -5 \leq x \leq 5\}$
b) $\{x \in \mathbb{Z} \mid x \leq 5\}$
c) $\mathbb{N} \cup \{0\}$
d) All of the above are countable sets.

Exercise 3.38. Prove that every subset of a countable set is countable.

Theorem 3.11. The set $\mathbb{Z}$ is countable. In particular, $|\mathbb{Z}| = \aleph_0$.

Proof. We show that the following function $f : \mathbb{Z} \to \mathbb{N}$ is a bijection.

$$f(n) = \begin{cases} 2n & \text{if } n > 0 \\ -2n + 1 & \text{if } n \leq 0 \end{cases} \tag{3.3}$$

Suppose that $f(n) = f(m)$. Since $f(n)$ is even if $n > 0$ and odd if $n \leq 0$, then either both $n, m > 0$ or both $n, m \leq 0$. In the first case, $2n = 2m$ and $n = m$. Similarly in the second case, $-2n + 1 = -2m + 1$ and $n = m$. This shows that f is one-to-one. To show onto, let $z \in \mathbb{N}$. If z is even then $f(z/2) = z$. If z is odd, then $f(\frac{z-1}{-2}) = z$. This shows that $f(\mathbb{Z}) = \mathbb{N}$. $\triangledown$

Exercise 3.39. Prove that the union of two countable sets is countable.

Exercise* 3.40. Suppose that $A_1, A_2, A_3, \ldots$ are countable sets. Prove that $\bigcup_{n \geq 1} A_n$ is again countable.

We have seen that the relation $\preceq$ is reflexive and transitive. The next paramount theorem on cardinal numbers states that $\preceq$ is anti-symmetric too, hence a partial order relation.

Theorem 3.12 (Cantor-Schröeder-Bernstein). For arbitrary sets A and B, if $|A| \preceq |B|$ and $|B| \preceq |A|$, then $|A| = |B|$.

Proof. Suppose that $f : A \to B$ and $g : B \to A$ are both injections. We shall construct a bijection in a rather sketchy manner as follows.

Let $A_1 = A$, $A_2 = g(B)$, and $A_n = g(f(A_{n-1}))$ for all $n \geq 3$. Observe, by induction, that $A_{n+1} \subseteq A_n$. Now let $S_n = A_n - A_{n+1}$ for $n \geq 1$, and let $S_{-1} = \bigcap_{n \geq 1} A_n$. It can be shown that A is partitioned into the subsets $S_{-1}, S_1, S_2, \ldots$, noting that S_n may actually be empty from some point on.

On the opposite side, let $B_0 = B$ and $B_n = f(A_n)$ for $n \geq 1$. Similarly, we have $B_{n+1} \subseteq B_n$, and let $T_n = B_n - B_{n+1}$ for $n \geq 0$. This time B is partitioned into $T_{-1}, T_0, T_1, \ldots$, where $T_{-1} = \bigcap_{n \geq 0} B_n$.

It is left as an exercise to show that $f(S_n) = T_n$ if n is odd, and $g(T_n) = S_{n+2}$ if n is even. The bijection we seek can now be constructed by putting together these "piecewise" bijections between the partitioning subsets. $\triangledown$

Theorem 3.13. The set $\mathbb{Q}$ is countable with $|\mathbb{Q}| = \aleph_0$.

Proof. We already have $|\mathbb{Q}| \succeq \aleph_0$. By Theorem 3.12, it suffices now to show $|\mathbb{Q}| \preceq \aleph_0$. Note that every rational number can be written m/n for some unique pair $(m, n) \in \mathbb{Z} \times \mathbb{N}$ with $\gcd(m, n) = 1$ (Exercise 2.59) if we agree to represent $0 = 0/1$. With the bijection f given in (3.3) earlier, the function $g(m/n) = (f(m), n)$ is clearly an injection $g : \mathbb{Q} \to \mathbb{N} \times \mathbb{N}$. In turn, the function $h(m, n) = 2^m \times 3^n$ is an injection $h : \mathbb{N} \times \mathbb{N} \to \mathbb{N}$, due to the uniqueness of prime factorization. The composition $h \circ g : \mathbb{Q} \to \mathbb{N}$ is then an injection which gives us $|\mathbb{Q}| \preceq |\mathbb{N}|$. $\triangledown$

Exercise 3.41. Prove that the cross product of two countable sets is again countable.

Our first example of a set that is not countable is the power set $P(\mathbb{N})$. By Theorem 3.12, it will suffice to show that $|P(\mathbb{N})| \succ \aleph_0$ for this claim. In fact, the cardinality of any set is always exceeded by that of its power set.

Theorem 3.14. For any set A, we have $|A| \prec |P(A)|$.

Proof. The function $f : A \to P(A)$ given by $f(a) = \{a\}$ is clearly an injection. To complete the proof, we show that if $g : A \to P(A)$ is any injection, then g is not onto. Simply let $S = \{a \in A \mid a \notin g(a)\} \in P(A)$. We claim that $g^{-1}(S) = \emptyset$, for if $g(a) = S$ for some $a \in A$, then $a \in S$ if and only if $a \notin g(a) = S$, a contradiction. Hence g cannot be onto. $\triangledown$

What is furthermore true is the fact that $\preceq$ is a total order relation on the set of cardinal numbers. We will not prove this claim, which depends on the so-called *well-ordering principle*, but we present a weaker version which suits our brief treatment on cardinal numbers.

Theorem 3.15. For any set A, exactly one of the relations $|A| \prec \aleph_0$ and $|A| = \aleph_0$ and $|A| \succ \aleph_0$ holds. In particular $|A| \prec \aleph_0$ if and only if A is finite.

Proof. Theorem 3.12 asserts that the three cases are mutually exclusive. It is obvious that there is an injection from any finite set A to $\mathbb{N}$, but never onto. Suppose now A is infinite; we will show that $|\mathbb{N}| \preceq |A|$. We claim that for every $n \in \mathbb{N}$, there is a subset $A_n \subseteq A$ with exactly n elements. Since $A \neq \emptyset$ this claim is true for $n = 1$. Using induction, we assume such A_n exists. Being infinite, $A \neq A_n$ so we can have $a \in A - A_n$. The subset $A_{n+1} = A_n \cup \{a\}$ completes the induction step.

Moreover, in the above construction we have $A_1 \subseteq A_2 \subseteq A_3 \subseteq \cdots \subseteq A$. Hence we have a subset $\{a_1, a_2, a_3, \ldots\} \subseteq A$ such that $a_k \in A_k$ and $a_k \neq a_j$ for all $j < k$. This gives the desired injection $f(n) = a_n$ from $\mathbb{N}$ to A. $\triangledown$

Definition. We call a set A *uncountable* if $|A| \succ \aleph_0$. Theorem 3.15 then asserts that being uncountable is indeed the negation of being countable. We also call A *countably infinite* when $|A| = \aleph_0$.

Hence, every set is either finite, countably infinite, or uncountable. In our definition, a countable set is either finite or countably infinite, but in other texts, the term countable is never used for finite sets.

For example, we have seen that both $\mathbb{Z}$ and $\mathbb{Q}$ are countably infinite sets, whereas $P(\mathbb{N})$ is uncountable. Another uncountable set is given by the real numbers.

Theorem 3.16. The set $\mathbb{R}$ is uncountable.

Proof. We will simply establish $|P(\mathbb{N})| \preceq |\mathbb{R}|$. Every $S \in P(\mathbb{N})$ corresponds to a unique infinite sequence $s_1, s_2, s_3, \ldots$, where $s_i = 1$ if $i \in S$ and $s_i = 0$ otherwise. (Recall the proof of Theorem 2.8.) Let $f : P(\mathbb{N}) \to \mathbb{R}$ be given by $f(s_1, s_2, s_3, \ldots) = \sum s_i \times 10^{-i}$, an infinite series convergent to a real number in the interval $[0, 1/9]$. For instance, $f(\emptyset) = 0$, $f(\mathbb{N}) = 0.\overline{1} = 1/9$, and $f(\{1, 4\}) = 0.1001$. We observe that f is one-to-one. $\triangledown$

Definition. We define $|\mathbb{R}| = c$, and call this quantity the *cardinality of the continuum.*

Exercise 3.42. Prove that the set of real numbers in the interval $(0, \infty)$ has cardinality c.

It can be shown that $|P(\mathbb{N})| = c$. Cantor's *continuum hypothesis* claims that there is no cardinal number strictly between $\aleph_0$ and c. Nevertheless, there are cardinal numbers larger than c, e.g., $c \prec |P(\mathbb{R})| \prec |P(P(\mathbb{R}))| \prec \cdots$ In 1963 Paul Cohen proved that the continuum hypothesis is independent of the common axioms of set theory.

3.4   Introduction to Groups

Group theory is normally taught as a first course in abstract algebra. Without going into much depth, we shall introduce the study of groups as a natural extension of sets and relations. In particular, we will be content being able to prove Fermat's little theorem, a promise we made upon stating Theorem 1.12.

Definition. By a *binary operation* $\star$ on a set S, we simply mean a function $f : S \times S \to S$ which is expressed by writing $f(a, b) = a \star b$. We say that $\star$ is *commutative* when $a \star b = b \star a$ for all $(a, b) \in S \times S$.

In practical example, a binary operation on $S = \mathbb{R}$ can be the ordinary addition or multiplication, or something like $a \star b = a + b + ab$, as long as the given function has its codomain in S.

Definition. A *group* G is a set together with a binary operation $\star$ on G which satisfies the following three conditions.

1) The operation $\star$ is *associative*, i.e., $a \star (b \star c) = (a \star b) \star c$ for all elements $a, b, c \in G$.

2) There exists an *identity element* $e \in G$, which has the property that $a \star e = e \star a = a$ for every element $a \in G$.

3) With a fixed identity element $e \in G$, each element $a \in G$ has an *inverse* $b \in G$, i.e., one for which $a \star b = b \star a = e$.

Example. We give several examples of what a group might look like.

1) The set of integers $\mathbb{Z}$ together with ordinary addition, which we know associative, is a group. The number $e = 0$ is an identity element, and an inverse of $a \in \mathbb{Z}$ is $-a$, since $a + (-a) = 0$. In fact, $\mathbb{Q}$ and $\mathbb{R}$, under addition, are also groups. From now on, when we say *the* group $\mathbb{Z}$, $\mathbb{Q}$, or $\mathbb{R}$, it is understood that the operation involved is addition.

2) The set of non-zero rational numbers $\mathbb{Q}^*$ is a group under the usual multiplication. Here, an identity element is $e = 1$, and each $a/b \in \mathbb{Q}^*$ has an inverse b/a. Similarly, the set $\mathbb{R}^* = \{x \in \mathbb{R} \mid x \neq 0\}$ is a group under multiplication. In the future, we refer to the groups $\mathbb{Q}^*$ and $\mathbb{R}^*$ without explicitly stating that they are under multiplication.

3) Note that $\mathbb{Z}^*$, the set of non-zero integers, is not a group under multiplication. If it were, the identity element would be $e = 1$. But then there will be no inverse for the element $2 \in \mathbb{Z}^*$.

4) The set $G = \{0\}$ under addition is a group, where 0 is the identity and only element in G. Essentially, this is the only kind of a group with one element, denoted by $\{e\}$, and it is called the *trivial group*.

5) The set $M(2, \mathbb{R})$ of 2×2 matrices with real entries is a group under matrix addition. Can you identify the identity element and inverses in this group? More generally, the set $M(n, S)$ of $n \times n$ matrices over S is a group under matrix addition, where S can be $\mathbb{Z}$, $\mathbb{Q}$, or $\mathbb{R}$.

6) The set $GL(2, \mathbb{R})$ of 2×2 matrices with non-zero determinants is also a group under matrix multiplication. We know from linear algebra that matrix multiplication is associative. The identity element here is $I = \begin{bmatrix} 1 & 0 \\ 0 & 1 \end{bmatrix}$, and recall that having a non-zero determinant is equivalent to being invertible.

Exercise 3.43. Let $G = \{x \in \mathbb{R} \mid x \neq -1\}$, and introduce the binary operation $a \star b = a + b + ab$ for all $a, b \in G$. Prove that G is a group.

Test 3.44. Which one of these four fails to be a group?

a) The set $\{3n \mid n \in \mathbb{Z}\}$ under addition.
b) The set $\{2^n \in \mathbb{Q} \mid n \in \mathbb{Z}\}$ under multiplication.
c) The set $\{a + b\sqrt{2} \mid a, b \in \mathbb{Z}\}$ under multiplication.
d) The set $\{A \in M(2, \mathbb{Z}) \mid \det A = \pm 1\}$ under matrix multiplication.

3.4.1 Some Basic Properties

Theorem 3.17. Every group has a unique identity element. Moreover, every element in a group has a unique inverse.

Proof. Let G be a group with two identity elements, e and f. Since e is identity, $e \star f = f$, and since f is identity, $e \star f = e$. Hence $e = f$. For the second claim, let $a \in G$ have two inverses b and c. Then $a \star b = a \star c = e$. Operate both sides by b from the left and then apply associativity to get

$$
\begin{aligned}
b \star (a \star b) &= b \star (a \star c) \\
(b \star a) \star b &= (b \star a) \star c \\
e \star b &= e \star c \\
b &= c
\end{aligned}
$$

This shows that a can have only one inverse.   ▽

Question. If $a \star b = a$ holds for two elements in a set S, can we conclude that S has an identity element, i.e., $e = b$?

From now on, we use the phrase *the* identity element, denoted by e, and *the* inverse of an element a, denoted by a^{-1}. Moreover, for convenience, we write ab instead of $a \star b$, unless sometimes when the operation is actually addition, then we write $a + b$ in order to avoid confusion. (Also, under addition, it is better to keep the notation for inverse as $-a$ instead of a^{-1}.) Due to associativity, we may then write the product abc without ambiguity, which generalizes to any finite number of elements, $a_1 a_2 a_3 \cdots a_n$, without the need of brackets.

Question. Is it true that $(a^{-1})^{-1} = a$ for any group element?

Exercise 3.45. Prove that $(ab)^{-1} = b^{-1} a^{-1}$ in any group.

Exercise* 3.46. If G and H are two groups, with their respective binary operations, prove that the direct product $G \times H = \{(g, h) \mid g \in G, h \in H\}$ is a group under the operation defined by $(g, h)(g', h') = (gg', hh')$.

Note that if a relation $ab = ac$ holds in a group, then $a^{-1}(ab) = a^{-1}(ac)$ and $b = c$ by associativity. This is a group property called the *cancellation law*, which may be applied left and right, as follows.

Theorem 3.18. Let G be a group and $a, b, c \in G$. If $ab = ac$ then $b = c$, and if $ba = ca$ then $b = c$.

Question. If $ab = a$ holds for two elements in a group G, can we conclude that the identity element of G is $e = b$?

As a precaution, we should not assume that cancellation laws always apply, unless we know that we are dealing with group elements. For example, we have $BA = CA$ with matrices

$$\begin{bmatrix} 0 & 1 \\ 0 & 2 \end{bmatrix} \begin{bmatrix} 1 & 2 \\ 2 & 4 \end{bmatrix} = \begin{bmatrix} 2 & 0 \\ 0 & 2 \end{bmatrix} \begin{bmatrix} 1 & 2 \\ 2 & 4 \end{bmatrix} = \begin{bmatrix} 2 & 4 \\ 4 & 8 \end{bmatrix}$$

but $B \neq C$, seemingly contradicting the theorem.

Question. What is wrong with the above false counter-example?

Definition. We say that a group G is *abelian* when the binary operation $\star$ on G is commutative. Also, we say that the group G is *finite* or *infinite* referring to G as a set.

All the examples that we have seen so far are abelian groups, except $GL(2, \mathbb{R})$, since matrix multiplication is not commutative.

Exercise 3.47. Prove that a group G is abelian if and only if, for all the elements of G, any one of the following propositions holds.

a) $ab = ca \rightarrow b = c$
b) $axb = cxd \rightarrow ab = cd$
c) $(ab)^2 = a^2b^2$
d) $(ab)^{-1} = a^{-1}b^{-1}$

Exercise* 3.48. Prove that if $a^2 = e$ for every element a in a group G with identity e, then G is abelian.

3.4.2 The Groups $\mathbb{Z}_n$ and $\mathbb{U}_n$

Until now we have not encountered any finite groups—other than the trivial group, of course—but soon we will see two families of groups which have a finite number of elements.

For a fixed positive integer n, let us define the set

$$\mathbb{Z}_n = \{0, 1, 2, \ldots, n-1\}$$

We have seen in Thereom 3.4 that the sets $[a]_n = \{a + nk \mid k \in \mathbb{Z}\}$, where $a \in \mathbb{Z}_n$, constitute the congruence (equivalence) classes under the relation $R = \{(a, b) \in \mathbb{Z} \times \mathbb{Z} \mid a \bmod n = b \bmod n\}$. Let us first claim some facts about these congruence classes.

Theorem 3.19. With $[a]_n = \{a + nk \mid k \in \mathbb{Z}\}$, suppose that $x \in [a]_n$ and $y \in [b]_n$. Then $x + y \in [a + b]_n$ and $xy \in [ab]_n$.

Question. How does this theorem compare with Theorem 1.11?

Proof. Let $x = a + nr$ and $y = b + ns$. Then $x + y = (a+b) + n(r+s) \in [a+b]_n$ and $xy = ab + n(as + br + nrs) \in [ab]_n$.   ▽

Definition. On the set $\mathbb{Z}_n$, we define the binary operations *addition mod* n and *multiplication mod* n by, respectively,

$$a +_n b = (a + b) \bmod n$$
$$a \times_n b = ab \bmod n$$

It is clear that both operations are commutative. To illustrate, we describe in tabular form the addition and multiplication mod 4 on the set $\mathbb{Z}_4 = \{0, 1, 2, 3\}$.

$+_4$	0	1	2	3
0	0	1	2	3
1	1	2	3	0
2	2	3	0	1
3	3	0	1	2

$\times_4$	0	1	2	3
0	0	0	0	0
1	0	1	2	3
2	0	2	0	2
3	0	3	2	1

Question. Can you see from the table why $\mathbb{Z}_4$ is *not* a group under $\times_4$?

Exercise 3.49. Construct the *multiplication* mod n table for $\mathbb{Z}_n$, for each $n = 5, 6, 7$, and 8. Ignoring the zero element, which tables suggest that we have a group?

We show now that $\mathbb{Z}_n$ is an abelian group under addition mod n.

1) By Theorem 3.19, $a +_n b \in [a + b]_n$. Hence, both $a +_n (b +_n c)$ and $(a +_n b) +_n c$ belong to $[a + b + c]_n$, and both lie in the range from 0 to $n - 1$. We conclude that, proving associativity,

$$a +_n (b +_n c) = (a +_n b) +_n c = (a + b + c) \bmod n$$

2) The identity element of $\mathbb{Z}_n$ is $e = 0$. This is obvious.

3) Every non-zero $a \in \mathbb{Z}_n$ has an inverse given by $-a = n - a$. This is true because $a +_n (n - a) = n \bmod n = 0$.

Exercise* 3.50. Show that $a \times_n (b +_n c) = (ab + ac) \bmod n$ for all elements $a, b, c \in \mathbb{Z}_n$.

We have hinted that in general $\mathbb{Z}_n$ is not a group under $\times_n$. Nevertheless, we will have a multiplicative group if we select only certain elements of $\mathbb{Z}_n$. More precisely, we consider the subset of $\mathbb{Z}_n$ defined by

$$\mathbb{U}_n = \{m \in \mathbb{Z}_n \mid \gcd(m, n) = 1\}$$

Take, for instance, $\mathbb{U}_{10} = \{1, 3, 7, 9\}$ with its multiplication table,

$\times_{10}$	1	3	7	9
1	1	3	7	9
3	3	9	1	7
7	7	1	9	3
9	9	7	3	1

Recall that $\gcd(m, n) = 1$ means that the two numbers have no common prime factor. Hence if $a, b \in \mathbb{U}_n$, then neither does ab have a common factor with n. In particular, we would have $1 = \gcd(ab, n) = \gcd(n, ab \bmod n)$ by Theorem 1.3. This shows that multiplication mod n is indeed a binary operation on $\mathbb{U}_n$.

Exercise 3.51. Construct the multiplication table for the set $\mathbb{U}_{12}$.

Question. If p is prime, what are the elements in $\mathbb{U}_p$?

To see why $\times_n$ is associative on $\mathbb{U}_n$, we similarly note that $a \times_n b \in [ab]_n$, so that we are able to conclude that

$$a \times_n (b \times_n c) = (a \times_n b) \times_n c = abc \bmod n$$

This time, the identity element of $\mathbb{U}_n$ is $e = 1$. Hence, for $\mathbb{U}_n$ to be a group, we are left to showing inverses. By Theorem 1.6, if $\gcd(m, n) = 1$ then we can find integers x and y such that $mx + ny = 1$. With these, $[mx]_n = [1]_n$ and so the inverse of m in $\mathbb{U}_n$ is given by $m^{-1} = x \bmod n$.

Exercise 3.52. Find the inverse of each element in the group $\mathbb{U}_{11}$.

We wrap up our introduction to these two finite groups by revealing their proper names.

Definition. Let $n \in \mathbb{N}$. The abelian group $\mathbb{Z}_n = \{0, 1, 2, \ldots, n-1\}$ under addition mod n is called the *group of integers mod n*. The elements $m \in \mathbb{Z}_n$ for which $\gcd(m, n) = 1$ are called *units*. And, the subset $\mathbb{U}_n = \{m \in \mathbb{Z}_n \mid \gcd(m, n) = 1\}$, which is an abelian group under multiplication mod n, is called the *group of units of $\mathbb{Z}_n$*.

3.4.3 Subgroups

Definition. A subset H of a group G is called a *subgroup* of G, in which case we write $H \subseteq G$, if H is itself a group under the same binary operation inherited from G.

Example. We illustrate the idea with several examples.

1) Since $\mathbb{Z} \subseteq \mathbb{Q} \subseteq \mathbb{R}$ as sets, and all three are groups under addition, we may state that $\mathbb{Z}$ is a subgroup of $\mathbb{Q}$, and that $\mathbb{Q}$ is a subgroup of $\mathbb{R}$.

2) The set $\mathbb{Q}^*$ under multiplication is a group and a subgroup of $\mathbb{R}^*$. The subset $\mathbb{Z}^*$ is not a subgroup of $\mathbb{Q}^*$, because $\mathbb{Z}^*$ is not a group under multiplication.

3) The subset $2\mathbb{Z}$ of even numbers is a subgroup of $\mathbb{Z}$ under addition. Note that the sum of two even numbers is again even, which says that addition is a binary operation on $2\mathbb{Z}$. That the three group axioms hold in $2\mathbb{Z}$ is easy to verify.

4) The set $S = \{1, -1\}$ forms a group under multiplication, so S is a subgroup of $\mathbb{Q}^*$. Although $\{1, -1\} \subseteq \mathbb{Z}^*$ as sets, we are not allowed to say that S is a subgroup of $\mathbb{Z}^*$, since the latter is not a group.

5) Every group is a subgroup of itself. Moreover, the trivial group $\{e\}$ can be viewed as a subgroup of every given group G.

6) The subset $\mathbb{U}_n$ is not a subgroup of $\mathbb{Z}_n$ even though both of them are groups, because they are defined with different binary operations.

7) The set $M(2, \mathbb{Z})$ is a subgroup of $M(2, \mathbb{R})$ under matrix addition.

Test 3.53. Which one of the following four sets is a subgroup of $\mathbb{Z}$?
a) $\{\pm 1\}$
b) $\{-1, 0, 1\}$
c) $\{6n \mid n \in \mathbb{Z}\}$
d) $\{2n + 1 \mid n \in \mathbb{Z}\}$

Exercise* 3.54. Prove that the group $GL(2, \mathbb{R})$ under matrix multiplication has a subgroup given by $SL(2, \mathbb{R})$, which consists of 2×2 matrices with determinant ± 1.

Observe that if $ab = a$ holds for two elements in a subgroup $H \subseteq G$, then by the cancellation law in G, we must have $b = e$. Similarly, if $ab = e$ then $b = a^{-1}$. We state these facts in the next theorem, followed by another theorem which serves as a subgroup test.

Theorem 3.20. If H is a subgroup of G, then the identity element of H is that of G. Moreover, for each $a \in H$, its inverse in H is the same $a^{-1} \in G$.

Exercise 3.55. If H and K are subgroups of G, prove that $H \cap K$ is also a subgroup of G. Conclude that the generalized intersection of any collection of subgroups is again a subgroup.

Exercise* 3.56. If H and K are subgroups of the same group G, is it true that $H \cup K$ is too a subgroup? Prove your claim or find a counter-example.

Theorem 3.21. A non-empty subset H of a group G is a subgroup if and only if $ab^{-1} \in H$ whenever $a, b \in H$.

Proof. Necessity is clear. Suppose the required condition is satisfied in H. Associativity in H is inherited from G. There is at least one element $a \in H$, hence $aa^{-1} = e \in H$. Also, for each $a \in H$ we have $ea^{-1} = a^{-1} \in H$. Last but not least, we have to verify that $a, b \in H$ implies $ab \in H$. This follows since $b \in H$ implies $b^{-1} \in H$, so $a, b \in H$ implies $a(b^{-1})^{-1} = ab \in H$. $\quad\triangledown$

Example. The set $n\mathbb{Z} = \{nk \mid k \in \mathbb{Z}\}$ under addition passes the subgroup test, since $nk + (-nj) = n(k - j) \in n\mathbb{Z}$. Thus $n\mathbb{Z} \subseteq \mathbb{Z}$. In particular, $2\mathbb{Z}$ is the subgroup of even numbers under addition.

Exercise* 3.57. Give a non-trivial example of a subgroup $H \subseteq \mathbb{Q}$ under addition, such that $\mathbb{Z} \subseteq H$.

It is not hard to see that Theorem 3.21 can also be presented as a two-step subgroup test: H is a subgroup if and only if (1) H is closed under multiplication, i.e., $a, b \in H \to ab \in H$ and (2) H is closed under inverse, i.e., $a \in H \to a^{-1} \in H$.

Exercise 3.58. Justify each subgroup relation $H \subseteq G$ claimed below.

a) $\{5n \mid n \in \mathbb{Z}\} \subseteq \mathbb{Z}$
b) $\{\pi^n \mid n \in \mathbb{Z}\} \subseteq \mathbb{R}^*$
c) $\{a + b\sqrt{2} \mid a, b \in \mathbb{Z} \wedge a^2 - 2b^2 = 1\} \subseteq \mathbb{R}^*$
d) $\{A \in M(2, \mathbb{Z}) \mid \det A = \pm 1\} \subseteq GL(2, \mathbb{R})$

Exercise 3.59. Let H be a *finite* non-empty subset of a group G. Show that H is a subgroup of G, if $ab \in H$ whenever $a, b \in H$.

Exercise* 3.60. The *centralizer* of an element a in a group G is defined by $C(a) = \{x \in G \mid ax = xa\}$. Show that $C(a) \subseteq G$, and conclude that the *center* of a group, i.e., $Z(G) = \{x \in G \mid ax = xa \,\forall a \in G\}$ is also a subgroup of G, upon observing that $Z(G) = \bigcap_{a \in G} C(a)$.

3.4.4 Cyclic Groups

Definition. Suppose that G is a group and $a \in G$. For every $k \in \mathbb{N}$, we define a^k recursively by $a^1 = a$ and $a^k = a^{k-1}a$. In addition, let $a^0 = e$ and $a^{-k} = (a^{-1})^k$.

Theorem 3.22. Let G be a group, $a \in G$, and $j, k \in \mathbb{Z}$. Then,

1) $a^{-k} = (a^{-1})^k = (a^k)^{-1}$

2) $a^j a^k = a^{j+k} = a^k a^j$

3) $(a^j)^k = a^{jk} = (a^k)^j$

Proof. If $k \geq 0$, then $a^{-k} a^k = (a^{-1})^k a^k = e$ by associativity. This holds for $k \leq 0$ as well by symmetry, proving that $a^{-k} = (a^k)^{-1}$. Also, if $k \leq 0$, $(a^{-1})^k = ((a^{-1})^{-1})^{-k} = a^{-k}$, proving (1). The rest is an exercise.   ▽

Exercise 3.61. Complete the proof of Theorem 3.22.

Question. Is it true that $a^k b^k = (ab)^k$?

Consistent with our earlier agreement, when the group operation is addition, we have $a^k = a + a + \cdots + a$. In that case, we choose to write ka in place of a^k. This also agrees with the fact that $0a = 0$, the identity element of addition. Similarly, $-ka = k(-a)$.

Definition. Let G be a group and $a \in G$. Let us define the set

$$\langle a \rangle = \{a^k \mid k \in \mathbb{Z}\}$$

and we will show that $\langle a \rangle$ is a subgroup of G, which we call the *cyclic subgroup of G generated by a*.

Theorem 3.23. For every $a \in G$, the set $\langle a \rangle$ is an abelian subgroup of G.

Proof. Commutativity is stated in Theorem 3.22, which also says that if $a^j, a^k \in \langle a \rangle$, then $a^j (a^k)^{-1} = a^j a^{-k} = a^{j-k} \in \langle a \rangle$. Hence, $\langle a \rangle$ passes the subgroup test of Theorem 3.21.   ▽

For example, the group $2\mathbb{Z}$ of even numbers is really the cyclic subgroup of $\mathbb{Z}$ generated by 2. And in general, we may write $n\mathbb{Z} = \langle n \rangle \subseteq \mathbb{Z}$.

Definition. Let G be a group and $a \in G$. If $\langle a \rangle = G$, then we call the group G *cyclic*. In such a case, then, G is generated by a, and so we call a a *generator* for the cyclic group G.

Question. Are all cyclic groups abelian? What about the converse?

Example. The group $\mathbb{Z}$ under addition is a cyclic group generated by 1. Similarly, $\mathbb{Z}_n = \langle 1 \rangle$ for all $n > 0$, under addition mod n. Another example is $\mathbb{U}_5 = \{1, 2, 3, 4\}$ under $\times_5$, where 2 and 3 are both generators.

Exercise 3.62. Find all the generators for the cyclic group $\mathbb{Z}_n$, for each $n = 8, 9, 10$, and 11.

Test 3.63. Which one of these four elements does *not* generate $\mathbb{Z}_{36}$?

a) 1
b) 15
c) 25
d) 35

Exercise 3.64. Find all the generators for the group $\mathbb{U}_n$, for each $n = 11$, 12, 13, and 14, if cyclic.

Unlike $\mathbb{Z}_n$, the group $\mathbb{U}_n$ is not always cyclic. In number theory, a generator for $\mathbb{U}_n$, if cyclic, is called a *primitive root* modulo n. It is known that primitive roots exist if and only if $n = 1, 2, 4, p^k$, or $2p^k$, for any prime $p > 2$ and $k \in \mathbb{N}$.

Exercise* 3.65. If a group has only three elements, prove that it must be cyclic.

Theorem 3.24. Any subgroup of a cyclic group is again cyclic.

Proof. Let $G = \langle a \rangle$ and $H \subseteq G$. If $H = \{e\}$ then $H = \langle e \rangle$, cyclic. Otherwise let $n \in \mathbb{N}$ be the least exponent for which $a^n \in H$. We claim that $H = \langle a^n \rangle$. Well, clearly $\langle a^n \rangle \subseteq H$. Now for each $a^m \in H$ we may write $m = qn + r$, with $q = \lfloor m/n \rfloor$ and $r = m \bmod n$. Then $a^r = a^m(a^{-n})^q \in H$. But since $0 \le r < n$, this would contradict the minimality of n, unless $r = 0$. Hence $a^m = (a^n)^q \in \langle a^n \rangle$, and it follows that $H \subseteq \langle a^n \rangle$. $\triangledown$

Example. Since $\mathbb{Z}$ is cyclic, every subgroup $H \subseteq \mathbb{Z}$ is given by $H = \langle n \rangle = n\mathbb{Z}$, i.e., the set of multiples of some $n \in \mathbb{Z}$. Moreover, as shown in the proof, n is the least positive integer in H. In particular, knowing that the intersection of subgroups is again a subgroup, we have $m\mathbb{Z} \cap n\mathbb{Z} = c\mathbb{Z}$, where c is the least natural number which is a multiple of both m and n. This leads us to the next result.

Theorem 3.25. If $m, n \in \mathbb{N}$ then $m\mathbb{Z} \cap n\mathbb{Z} = \mathrm{lcm}(m, n)\mathbb{Z}$. In particular, if $\gcd(m, n) = 1$, then $m\mathbb{Z} \cap n\mathbb{Z} = mn\mathbb{Z}$.

Exercise* 3.66. Suppose that G and H are both cyclic groups. Give an example where $G \times H$ is cyclic, and another where $G \times H$ is not cyclic.

3.4.5 Cosets

Definition. Let H be a subgroup of a group G. For elements $a, b \in G$, we define the relation $a \sim b$ if and only if $ab^{-1} \in H$.

Your job is to prove that $a \sim b$ defines an equivalence relation on G. For example, if $G = \mathbb{Z}$, under addition, and $H = \langle n \rangle$, then $a \sim b$ if and only if $a - b \in \langle n \rangle$. But this is just the congruence relation $a \bmod n = b \bmod n$, which we proved in Theorem 3.4.

Exercise 3.67. Given $H \subseteq G$, verify that $R = \{(a, b) \mid a \sim b\}$ is indeed an equivalence relation on G, where $a \sim b$ if and only if $ab^{-1} \in H$.

Definition. We call the equivalence class of $a \in G$ under this relation the *coset* of a in G with respect to the subgroup H, given by

$$
\begin{aligned}
Ha &= \{b \in G \mid b \sim a\} \\
&= \{b \in G \mid ba^{-1} \in H\} \\
&= \{b \in G \mid ba^{-1} = h \wedge h \in H\} \\
&= \{b \in G \mid b = ha \wedge h \in H\} \\
&= \{ha \mid h \in H\}
\end{aligned}
$$

Hence, for example, with the relation $a \equiv b \pmod{n}$ on $\mathbb{Z}$, where $H = \langle n \rangle$, the cosets are the congruence classes $[a]_n$, in agreement with the fact that $Ha = \{h + a \mid h \in \langle n \rangle\} = \{nk + a \mid k \in \mathbb{Z}\} = [a]_n$.

Example. Consider $\mathbb{U}_7 = \{1, 2, 3, 4, 5, 6\}$ with its subgroup $\langle 2 \rangle = \{1, 2, 4\}$. For each element $a \in \mathbb{U}_7$, we compute the coset $\langle 2 \rangle a$:

$$
\begin{array}{lll}
\langle 2 \rangle 1 = \{1, 2, 4\} & \langle 2 \rangle 2 = \{2, 4, 1\} & \langle 2 \rangle 3 = \{3, 6, 5\} \\
\langle 2 \rangle 4 = \{4, 1, 2\} & \langle 2 \rangle 5 = \{5, 3, 6\} & \langle 2 \rangle 6 = \{6, 5, 3\}
\end{array}
$$

Note that in this example, only two of the cosets are distinct.

Exercise 3.68. Describe the cosets of each given group G, induced by the relation $a \sim b$, with respect to the subgroup $H \subseteq G$.
a) $\langle 18 \rangle \subseteq \mathbb{Z}_{24}$
b) $\langle 3 \rangle \subseteq \mathbb{U}_{13}$
c) $5\mathbb{Z} \subseteq \mathbb{Z}$
d) $\{\pm 1\} \subseteq \mathbb{Q}^*$

Definition. The number of distinct cosets, with respect to the subgroup $H \subseteq G$ and the relation $a \sim b$, is denoted by $[G{:}H]$. We call this quantity $[G{:}H]$ the *index* of H in G, which could happen to be infinite. Moreover, let us call $|G|$, i.e., the cardinality of the set G, the *order* of the group G.

From the properties of equivalence classes, we conclude that these cosets form a partition for the group G. From this fact, we now deduce a series of results leading to our goal of proving Fermat's little theorem.

Theorem 3.26. Let G be any group with a subgroup H. For each $a \in G$, we have $|Ha| = |H|$.

Proof. Every element in Ha is of the form ha for some $h \in H$. Moreover, $ha = h'a$ implies $h = h'$ by the cancellation law. This gives a one-to-one correspondence between Ha and H which proves the claim, regardless H is finite or infinite. $\triangledown$

Theorem 3.27 (Lagrange's Theorem). The order of any subgroup H divides the order of the group G, provided that G is finite. In such a case, $|G|/|H| = [G{:}H]$.

Proof. Let G be a finite group and $H \subseteq G$. There can be only finitely many cosets in G with respect to H, say $[G{:}H] = k$. Since G is partitioned into the k cosets, we have $k|H| = |G|$ by Theorem 3.26. $\triangledown$

Exercise 3.69. Suppose that H and K are finite subgroups of G. If $\gcd(|H|, |K|) = 1$, show that $H \cap K$ is the trivial group.

Exercise* 3.70. Show that any group of prime order is cyclic and, in such a case, any non-identity element is a generator.

Definition. Let G be a group and $a \in G$. The *order* of a in G, denoted by $|a|$, is the smallest $n \in \mathbb{N}$ such that $a^n = e$. If there is no such number n, we let $|a| = |\langle a \rangle|$.

For example, in $\mathbb{U}_5$ we have $2^2 = 4$, $2^3 = 3$, and $2^4 = 1$; hence $|2| = 4$ in this group. Similarly, in $\mathbb{Z}_{12}$ we have $|2| = 6$. The next theorem explains why we choose the cardinal number $|a| = |\langle a \rangle|$ for an alternative.

Exercise 3.71. Suppose that $|a| = 12$ in the group G. Determine the order of $a^k \in G$, for each k in the range $1 \leq k \leq 12$.

Test 3.72. Given that $|a| = 24$ in G, what is the order of $a^{15} \in G$?
a) 8
b) 24
c) 72
d) 120

Theorem 3.28. Let a be an element of a group G. Then $|a| = |\langle a \rangle|$.

Proof. Let $|a| = n \in \mathbb{N}$ (else nothing to prove), and let $H = \{a, a^2, \ldots, a^n\}$. Note that $a^n = e$. First, we claim that $|H| = n$ by showing that the elements $a, a^2, \ldots, a^n$ are all distinct. To see why, suppose $a^j = a^k$ with $1 \leq j < k \leq n$. Then $a^{k-j} = e$ with $0 < k - j < n$, contradicting the minimality of n. Next, we would be done if $H = \langle a \rangle$. Clearly, $H \subseteq \langle a \rangle$. Now let $a^m \in \langle a \rangle$. We write $m = qn+r$, where $q = \lfloor m/n \rfloor$ and $r = m \bmod n$. Since $a^n = e$, then $a^m = (a^n)^q a^r = a^r \in H$ as $0 \leq r < n$. Thus $\langle a \rangle \subseteq H$. $\triangledown$

Theorem 3.29. The order of any element in a finite group divides the order of the group. As a consequence, for every $a \in G$, we have $a^{|G|} = e$.

Proof. For finite groups, $|a| = |\langle a \rangle| = n \in \mathbb{N}$, and by Lagrange's theorem, this quantity divides $|G|$. Hence, $a^{|G|} = (a^n)^{|G|/n} = e$. $\quad \triangledown$

And now, if p is a prime number, and if we let $a \in \mathbb{U}_p = \{1, 2, \ldots, p-1\}$ in Theorem 3.29, then $a^{p-1} \bmod p = 1$. With a touch from Theorem 3.19, or 1.11, Fermat's little theorem follows at last.

Theorem 3.30 (Fermat's Little Theorem). Let $a \in \mathbb{Z}$ and p be a prime, not dividing a. Then $a^{p-1} \bmod p = 1$. More generally, if $\gcd(a, n) = 1$ then $a^{\phi(n)} \bmod n = 1$, where $\phi(n) = |\mathbb{U}_n|$.

Definition. The function $\phi(n) = |\mathbb{U}_n|$, with domain $\mathbb{N}$, is called the *Euler's phi function*, and the generalization of Fermat's little theorem for $\mathbb{U}_n$ is better known as Euler's theorem.

Exercise* 3.73. Evaluate $\phi(p^n)$, where p is a prime number and $n \geq 1$.

3.4.6 Finite Cyclic Groups

Given a cyclic group $G = \langle a \rangle$ of finite order n, we seek to identify the order of each element $a^k \in G$, where $1 \leq k \leq n$. Since every subgroup of G is generated by one element, as G itself is, such knowledge will lead to the classification of all the subgroups of G.

Theorem 3.31. Let $a \in G$, not assumed cyclic. Then $a^k = e$ if and only if $|a|$ divides k.

Proof. Let $|a| = n$ and write $k = qn + r$ with $0 \leq r < n$. We have $a^k = (a^n)^q a^r = a^r$. By the minimality of n, then $a^k = e$ if and only if $r = 0$. $\quad \triangledown$

Exercise 3.74. Let G and H be two finite cyclic groups. Show that $G \times H$ is again cyclic, if $\gcd(|G|, |H|) = 1$.

Exercise* 3.75. If $\gcd(m, n) > 1$, prove that $\mathbb{Z}_m \times \mathbb{Z}_n$ is *not* cyclic.

Theorem 3.32. Suppose $a \in G$, not assumed cyclic, such that $|a| = n$. Then $|a^m| = n/\gcd(m, n)$.

Proof. Assume first $\gcd(m, n) = 1$. Let $|a^m| = k$. Then $a^{mk} = e$ and n divides mk by Theorem 3.31. But n has no common factor with m, hence n must divide k. In particular, $k \geq n$. Since $(a^m)^n = (a^n)^m = e$, the minimality of k implies that $k = n$. Thus $|a^m| = n/\gcd(m, n)$.

Now suppose $\gcd(m, n) = d > 1$. Note that $s = n/d$ is the least natural number for which $(a^d)^s = e$. Thus $|a^d| = n/d$. Since $\gcd(m/d, n/d) = 1$, the same argument above gives $|a^m| = |(a^d)^{m/d}| = n/d = n/\gcd(m, n)$. ∇

Test 3.76. Given that $|a| = 12$ in G, which one of the following four elements also has order 12?

a) a^{10}
b) a^{12}
c) a^{25}
d) a^{27}

Theorem 3.33. Let G be a cyclic group of order n, generated by a. Let $d \in \mathbb{N}$ be any divisor of n. Then,

1) $G = \langle a^m \rangle$ if and only if $\gcd(m, n) = 1$.

2) G has exactly $\phi(n)$ generators.

3) G has exactly $\phi(d)$ elements of order d.

4) G has a unique subgroup of order d.

Proof. We have $\langle a^m \rangle = \langle a \rangle$ if and only if $|a^m| = |a|$, i.e., if and only if $\gcd(m, n) = 1$, according to Theorem 3.32. Writing $G = \{e, a, a^2, \ldots, a^{n-1}\}$, we see that $G = \langle a^m \rangle$ if and only if $m \in \mathbb{U}_n$. Hence, G has $\phi(n)$ generators. This proves the first two claims.

Since $|a^{n/d}| = d$, we have a subgroup $\langle a^{n/d} \rangle$ of order d. To show uniqueness, suppose also $|\langle a^k \rangle| = d = n/\gcd(k, n)$. Then $\gcd(k, n) = n/d$. In particular, n/d divides k, and $a^k \in \langle a^{n/d} \rangle$. Hence $\langle a^k \rangle \subseteq \langle a^{n/d} \rangle$ and, being of the same size, we conclude that $\langle a^k \rangle = \langle a^{n/d} \rangle$, proving (4).

It also follows that every element b of order d will give us $\langle b \rangle = \langle a^{n/d} \rangle$. In particular, b generates the cyclic group of order d, and we have shown that there are $\phi(d)$ such generators. ∇

Exercise 3.77. Let $m \in \mathbb{Z}_n$. Prove that $\mathbb{Z}_n = \langle m \rangle$ if and only if $m \in \mathbb{U}_n$.

Test 3.78. Given that $\mathbb{U}_{19}$ is cyclic, how many generators does it have?

a) 1
b) 6
c) 9
d) 18

Exercise* 3.79. Prove that $\mathbb{U}_n$ has exactly $\phi(\phi(n))$ generators, if cyclic.

Test 3.80. If G is cyclic and $|G| = 96$, how many elements have order 24?

a) 0
b) 4
c) 8
d) 12

Exercise* 3.81. For every $n \in \mathbb{N}$, show that $\sum \phi(d) = n$, where d ranges over all the divisors of n.

Combined with Lagrange's theorem, Theorem 3.33 yields a one-to-one correspondence between subgroups of a cyclic group G and divisors of $|G|$. In fact, the subgroups are themselves cyclic, so for any two of them, the relation $H \subseteq K$ holds if and only if $|H|$ divides $|K|$.

And now, since *divisor-of* is a partial order relation on $|G|$, its Hasse diagram corresponds to that for the subgroups of G under the *subgroup-of* relation, which we call the *subgroup lattice* of the finite cyclic group G.

Example. To draw the subgroup lattice of $\mathbb{Z}_{12}$, we start with the set of positive divisors of 12, i.e., $A = \{1, 2, 3, 4, 6, 12\}$. For each $d \in A$, we identify the unique subgroup of order d—in this case $\langle 12/d \rangle$. Both Hasse diagrams, for $R = \{(a, b) \mid b \mod a = 0\}$ on A and for $S = \{(H, K) \mid H \subseteq K\}$ on the subgroups of G are compared side-by-side below.

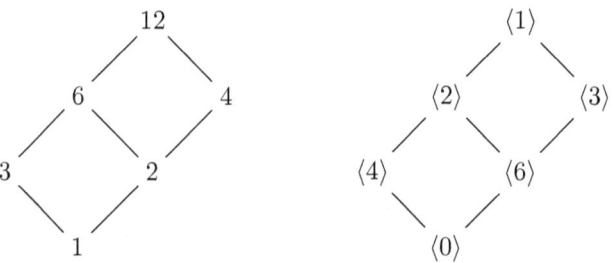

Exercise 3.82. Draw the subgroup lattice for each of the cyclic groups $\mathbb{Z}_{30}$, $\mathbb{Z}_{36}$, and $\mathbb{U}_{17}$.

3.4.7   Permutation Groups

To conclude this section, we shall introduce an entirely different kind of groups, in the sense that they are not number sets. In fact, we will have a group of bijective functions on a finite set, otherwise known as permutations, and in particular, a family of permutation groups related to geometry.

Definition. By a *permutation* on a set A, we mean a function $f : A \rightarrow A$ which is one-to-one and onto. For $n \in \mathbb{N}$, we denote by S_n the set of all permutations on the set $\{1, 2, 3, \ldots, n\}$. It is not hard to see that $|S_n| = n!$ and that it is a group under function composition. We call S_n the *symmetric group* of degree n and call any subgroup of S_n a *permutation group*.

Exercise 3.83. Verify that S_n is a group of order $n!$ under composition.

Example. Consider S_6, the set of $6! = 720$ permutations on $\{1, 2, 3, 4, 5, 6\}$. An element $f \in S_6$ may be written in *cyclic notation*; e.g., $f = (1, 2, 5)(3, 6)$, which means that the function f is given by

$$f(1) = 2 \qquad f(2) = 5 \qquad f(3) = 6$$
$$f(4) = 4 \qquad f(5) = 1 \qquad f(6) = 3$$

Note that 4 is missing in the notation; this is understood as $f(4) = 4$. In general, a number left unchanged by the permutation can be omitted from the cyclic notation, except when writing the identity permutation, i.e., $e = (1)$.

Definition. The term *cycle* refers to each bracketed part in the cyclic notation. It is intuitively clear that every permutation can be represented by *disjoint* cycles, i.e., where no two cycles contain a common term.

For example, the permutation $(1, 2, 5)(3, 6)$ is written in two disjoint cycles. Moreover, the terms in a cycle may be permuted in a circular manner, e.g., $(1, 2, 5) = (2, 5, 1) = (5, 1, 2)$.

Following convention, recall that composition is read from right to left, and it is generally non-commutative, e.g.,

$$(1, 2, 5)(3, 6) \circ (4, 6, 2, 1) = (1, 4, 3, 6, 5)$$
$$(4, 6, 2, 1) \circ (1, 2, 5)(3, 6) = (2, 5, 4, 6, 3)$$

But note that disjoint cycles commute: e.g., $(1, 2, 5) \circ (3, 6) = (1, 2, 5)(3, 6) = (3, 6) \circ (1, 2, 5)$.

Exercise 3.84. Show that S_n is non-abelian for all $n \geq 3$.

Exercise 3.85. Determine the order of each element of S_n given below.

a) (2,1,5,6,4,3)
b) (2,1,5)(6,4,3)
c) (2,1,5)(6,4)
d) (2,1,5)(6,4)(3,9,7,8)

Exercise* 3.86. Draw the subgroup lattice for $\langle f \rangle$, a cyclic subgroup of S_5 generated by $f = (1, 2, 3)(4, 5)$.

Consider a regular n-gon whose vertices are labeled 1 to n in a counter-clockwise sequence. There are n symmetries which are obtained by rotations around the center. It is clear that these n rotations form a cyclic subgroup $\langle R \rangle$ of S_n, where $R = (1, 2, 3, \ldots, n)$, i.e., the $2\pi/n$ rotation.

Moreover, there are n symmetries which result from reflections across the n axes of symmetry. Note that $|F| = 2$ for each reflection $F \in S_n$. These $2n$ rotations and reflections form a permutation group, which is called the *dihedral group* of degree n, denoted by D_n.

Example. With $n = 4$, there are four rotations, $R = (1, 2, 3, 4), R^2, R^3$, and $R^4 = e$, which correspond to $\pi/2, \pi, 3\pi/2$, and 2π radians, respectively.

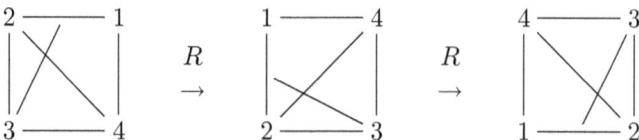

The four reflections can be represented by $F_1 = (1, 4)(2, 3)$, $F_2 = (2, 4)$, $F_3 = (1, 2)(3, 4)$, and $F_4 = (1, 3)$—symmetries with respect to the x-axis, the line $y = x$, the y-axis, and the line $y = -x$, respectively.

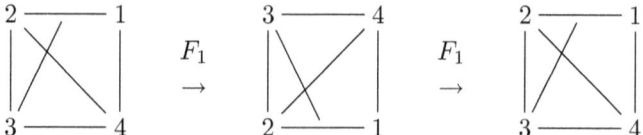

The claim that D_4 is a group is supported by its composition table below, keeping in mind that composition reads right to left.

Table 3.1: The composition table of the group D_4.

$\circ$	R	R^2	R^3	R^4	F_1	F_2	F_3	F_4
R	R^2	R^3	R^4	R	F_2	F_3	F_4	F_1
R^2	R^3	R^4	R	R^2	F_3	F_4	F_1	F_2
R^3	R^4	R	R^2	R^3	F_4	F_1	F_2	F_3
R^4	R	R^2	R^3	R^4	F_1	F_2	F_3	F_4
F_1	F_4	F_3	F_2	F_1	R^4	R^3	R^2	R
F_2	F_1	F_4	F_3	F_2	R	R^4	R^3	R^2
F_3	F_2	F_1	F_4	F_3	R^2	R	R^4	R^3
F_4	F_3	F_2	F_1	F_4	R^3	R^2	R	R^4

Exercise* 3.87. Describe the cosets of S_4 with respect to the subgroup D_4 and the relation $a \sim b$ as before.

Exercise 3.88. Construct the composition tables for D_3 and D_5, clearly distinguishing between the rotations and the reflections.

Being a finite subset of S_n, by Exercise 3.59, D_n is a subgroup if and only if it is closed under composition, meaning that $g \circ f \in D_n$ whenever $f, g \in D_n$. The key facts needed in the proof are summarized in the next two exercises.

Exercise 3.89. A rotation shifts the vertices of the n-gon but preserves their counter-clockwise circular ordering. A reflection, on the other hand, reverses the ordering in the clockwise direction. Prove these claims.

Exercise 3.90. In D_n, show that the composition of two rotations or two reflections is a rotation, whereas a mixed composition yields a reflection.

Exercise 3.91. Determine the order of each element in D_n.

Since $|D_n| = 2n$, in particular we have $D_3 = S_3$. Other than this exception, D_n is a proper subgroup of S_n and is non-abelian as S_n is.

Exercise 3.92. If $F \in D_n$ is a reflection, show that $F \circ R = R^{n-1} \circ F$, and conclude that D_n is non-abelian for all $n \geq 3$.

Exercise* 3.93. Prove that a subgroup of D_n is cyclic, if its order is odd.

Books to Read

1. E. D. Bolker, *Elementary Number Theory: An Algebraic Approach*, 1970, Dover Publications 2007.

2. P. J. Cohen, *Set Theory and the Continuum Hypothesis*, 1966, Dover Publications 2008.

3. J. F. Humphreys and M. Y. Prest, *Numbers, Groups and Codes*, Second Edition, Cambridge University Press 2004.

4. D. Joyner, *Adventures in Group Theory: Rubik's Cube, Merlin's Machine, and Other Mathematical Toys*, Second Edition, Johns Hopkins University Press 2008.

Chapter 4

Topics in Combinatorics

Combinatorics deals with discrete structures that are governed by certain patterns with regard to the way these structures are arranged. Terms like permutation, ordering, partition, sequence, are words that one may encounter in a combinatorial problem, just to name a few examples. In this chapter, we select mainly topics which are related to the theory of counting.

4.1 Techniques of Counting

Loosely speaking, in counting theory we ask questions which start with *how many?* Essentially, this will translate to finding the cardinality of a set A, where A consists of the elements which obey certain patterns under consideration. We shall introduce a few preliminary counting techniques leading to the familiar notion of permutations and combinations.

4.1.1 Some Basic Principles

Two foundational counting principles are as basic as addition and multiplication are in arithmetic. The *addition principle* is a realization of the fact that if A is partitioned into subsets $A_1, A_2, \ldots, A_n$, then $|A| = \sum |A_i|$. Meanwhile, the *multiplication principle* applies in counting the number of ordered pairs $(a, b) \in A \times B$, where the formula $|A \times B| = |A| \times |B|$ is known.

To understand these two principles, we just have to see how they apply in different counting situations.

Example. How many two-digit integers have their digit sum at most 4?

Solution. Clearly, the first digit will have to be 1, 2, 3, or 4. The first case applies to 10, 11, 12, and 13—four of them. Similarly, there are three, two and one, respectively for the remaining cases. By the addition principle, the answer is then $4 + 3 + 2 + 1 = 10$ numbers.

Exercise 4.1. How many three-digit integers have their digits sum to 7?

Example. An obsolete Zimbabwean banknote denomination was printed with a serial number consisting of 2 letters of the alphabet, followed by a string of seven digits between 0 and 9 each. How many of such banknotes could have been printed without exhausting the serial numbers?

Solution. The multiplication principle applies here. With nine elements, 26 choices (letters) for the first two and 10 choices (digits) for the remaining seven, we get a total of $26 \times 26 \times 10^7 = 6,760,000,000$ serial numbers.

Exercise 4.2. At Planet University, students are assigned their identification numbers with a string of 9 digits. The first four are reserved for the year in which the student first registers, followed by 1, 2, or 3, to indicate the semester. The sixth digit, between 1 and 8, is for Faculty code, and the last three for the student's serial number within the Faculty, from 001 to 999. How many ID numbers can be assigned in 20 years?

Exercise* 4.3. How many numbers between 10,000 and 100,000 are odd and consist of distinct digits? Hint: assigning the digits from left to right is not the best approach.

Example. How many integers up to 1000 contain a unique digit of 7?

Solution. We consider numbers between 000 to 999. If 7 is the first digit, there are nine digits to choose from for the second digit, as well as the third. By the multiplication principle, there are $9 \times 9 = 81$ such numbers. The same figure we have if 7 is the second, or third, digit. By the addition principle, we have $81 + 81 + 81 = 243$ in all.

Exercise 4.4. How many five-digit numbers, no less than 47000, can have no repeated digits?

Exercise* 4.5. Let c_n denote the number of *compositions*, or ordered partitions, of n into positive integers. For example, $c_4 = 8$, since there are $4 = 3 + 1 = 2 + 2 = 2 + 1 + 1 = 1 + 3 = 1 + 2 + 1 = 1 + 1 + 2 = 1 + 1 + 1 + 1$. Prove that $c_n = 2^{n-1}$ for all $n \geq 1$.

Example. How many positive integers are divisors of 600?

Solution. We have the prime factorization $600 = 2^3 \times 3 \times 5^2$. By the fundamental theorem of arithmetic, a divisor d must factor into the same primes, each with equal exponent or less. Hence, $d = 2^a \times 3^b \times 5^c$, where $a \in \{0, 1, 2, 3\}$, $b \in \{0, 1\}$, and $c \in \{0, 1, 2\}$. By the multiplication principle, there are $4 \times 2 \times 3 = 24$ such divisors.

Exercise 4.6. Count how many positive divisors each number has.
a) 777
b) 2010
c) 99000
d) 12321

Another useful principle states that if k sets have $k + 1$ elements in all, then one of the k sets, perhaps more, must contain at least two elements. For instance, in any group of 13 people, at least two will have their birthdays in the same month—quite easy to conclude since a year has only 12 months. This principle is known by the name of the *pigeonhole principle,* two of whose generalizations will be given after we see some applications.

Example. Choose any three integers. Prove that we can find two of them whose sum is even.

Solution. An integer must be even or odd; call these two *parity* classes. With three integers, two must belong to the same parity: both odd or both even, and their sum will be even.

Example. Consider a sequence of n integers $a_1, a_2, \ldots, a_n$. Prove that there exist consecutive terms in this sequence whose sum is divisible by n.

Solution. Let $s_0 = 0$, $s_1 = a_1$, $s_2 = a_1 + a_2$, $\ldots$, $s_n = a_1 + a_2 + \cdots + a_n$. A remainder upon dividing an integer by n lies in the range from 0 to $n - 1$. With $n + 1$ remainders—s_0 mod n, s_1 mod n, $\ldots$, s_n mod n—two of them must be identical, say s_i and s_{i+j}. The difference $s_{i+j} - s_i$ will then be a multiple of n, and note that $s_{i+j} - s_i = a_{i+1} + a_{i+2} + \cdots + a_{i+j}$.

Exercise 4.7. Let 1001 integers be chosen between 1 and 2000 inclusive. Use the pigeonhole principle to prove that one integer in the selection is a multiple of another one, by showing that two of them have the same largest odd factor.

Theorem 4.1 (Pigeonhole Principle, First Form). If k sets have n elements in all, then one of them must contain at least $\lceil n/k \rceil$ elements.

Proof. Let m be the largest cardinality among the k sets. To keep m at minimum, the n elements must be equally distributed as much as possible. This can occur when n/k is an integer, in which case $m = n/k$, else some sets will contain $\lceil n/k \rceil$ elements. $\triangledown$

Exercise 4.8. There are seven chapters to study for a Calculus exam. What is the minimum number of questions in the exam in order to guarantee that five of them are taken from the same chapter?

Example. Let six arbitrary sets be given. Show that there exist three of them which are either mutually disjoint or mutually overlapping.

Solution. Let A be one of the six sets. Separate the remaining five into two groups, depending whether the set overlaps or is disjoint with A. By the pigeonhole principle, there exist three which belong together in the same group. If these three are mutually disjoint or mutually overlapping, we are done. Otherwise, we can find two of the three which are disjoint and two which are overlapping. Whichever group it is, choose the pair which, together with A, they make mutually disjoint or mutually overlapping three.

Theorem 4.2 (Pigeonhole Principle, Second Form). Let $n_1, n_2, \ldots, n_k$ be positive integers whose sum is n. If $A_1, A_2, \ldots, A_k$ are sets with a total of $n - k + 1$ elements, then $|A_i| \geq n_i$ for at least one of the k sets.

Proof. If the claim were false, $|A_i| \leq n_i - 1$ for each of them, and together there would be at most $n - k$ elements in all, a contradiction. $\triangledown$

Example. Philadelphia Donuts offers four kinds of donuts: sugar, eclair, blueberry, and lemon. How many donuts, at minimum, does Elias need to buy in order to guarantee that there will be three of a kind? (For some reason, Elias leaves it to the waiter to select any combination.)

Solution. According to the pigeonhole principle, it suffices to buy $3 + 3 + 3 + 3 - 4 + 1 = 9$ donuts. Note that, if only 8 donuts, bad luck might give Elias two of each kind.

4.1.2 The Inclusion-Exclusion Principles

The principles of inclusion-exclusion apply in counting the elements in the union of a number of sets. The simplest principle, involving just two sets, is the fact that in counting the members in $A \cup B$, those in $A \cap B$ contribute to $|A|$ as well as to $|B|$. The result is a formula stated as Theorem 4.3, to be succeeded by its generalization for three sets.

Theorem 4.3. For any two sets, $|A \cup B| = |A| + |B| - |A \cap B|$.

Example. How many integers from 1 to 100 are divisible by 4 or 6?

Solution. Let A consist of the multiples of 4 in the range, and similarly B of 6. Since every fourth number belongs to A, then $|A| = 100/4 = 25$. Similarly, $|B| = \lfloor 100/6 \rfloor = 16$. (Note that the floor function is needed in order to get the right integer value.) Now according to Theorem 3.25, $A \cap B$ consists of the multiples of $\operatorname{lcm}(4,6) = 12$, hence $|A \cap B| = \lfloor 100/12 \rfloor = 8$. The final answer is therefore, $25 + 16 - 8 = 33$.

Question. How many integers up to 100 are divisible by *neither* 4 nor 6?

Exercise 4.9. Count how many integers from 1 to 1000 which satisfy the given conditions.
a) Multiples of 11 or 13.
b) Divisible by 28 or 40.
c) Divisible by neither 28 nor 40.
d) Not multiples of 27 or of 45.

By the way, though not an inclusion-exclusion principle, the fact that A is partitioned into $A - B$ and $A \cap B$ gives us the following related formula.

$$|A - B| = |A| - |A \cap B|$$

Test 4.10. From 1 to 1000, how many are divisible by 30 but not by 18?
a) 22
b) 32
c) 43
d) 55

Theorem 4.4. Let A, B, C be three sets. Then,

$$|A \cup B \cup C| = |A| + |B| + |C| - |A \cap B| - |A \cap C| - |B \cap C| + |A \cap B \cap C|$$

Proof. By Theorem 4.3, $|A \cup B \cup C| = |A| + |B \cup C| - |A \cap (B \cup C)|$ and $|B \cup C| = |B| + |C| - |B \cap C|$. We are left with showing that

$$|A \cap (B \cup C)| = |A \cap B| + |A \cap C| - |A \cap B \cap C|$$

This follows after the identity $A \cap (B \cup C) = (A \cap B) \cup (A \cap C)$ of Theorem 2.6 and one more application of Theorem 4.3. $\triangledown$

Example. How many integers from 1 to 200 are divisible by 4, 6, or 10?

Solution. Denote by A, B, C, respectively, the multiples of 4, 6, 10, in the given range. Noting that $\operatorname{lcm}(4,6) = 12$, $\operatorname{lcm}(4,10) = 20$, $\operatorname{lcm}(6,10) = 30$,

$$|A| = 200/4 = 50 \qquad\qquad |A \cap B| = \lfloor 200/12 \rfloor = 16$$
$$|B| = \lfloor 200/6 \rfloor = 33 \qquad\qquad |A \cap C| = 200/20 = 10$$
$$|C| = 200/10 = 20 \qquad\qquad |B \cap C| = \lfloor 200/30 \rfloor = 6$$

Furthermore, $\mathrm{lcm}(4, 6, 10) = 60$ and $|A \cap B \cap C| = \lfloor 200/60 \rfloor = 3$, giving the final answer $50 + 33 + 20 - 16 - 10 - 6 + 3 = 74$.

Question. What is the best way to evaluate $\mathrm{lcm}(a, b, c)$?

Exercise 4.11. Count how many integers from 1 to 1000 with the given conditions.

a) Multiples of 11, 13 or 15.
b) Divisible by 28, 35, or 40.
c) Divisible by neither 28, 35, nor 40.
d) Not multiples of 27, 45, or 54.

Exercise* 4.12. State and prove the inclusion-exclusion principle for four sets.

Example. If we compute the number 123! we will see that it ends with several zeros. Count the number of terminating zeros in 123!.

Solution. We count the powers of 10 in it. Since $10 = 2 \times 5$, it suffices to count the multiples of 5 or 5^2, since the multiples of 2 clearly outnumber them. However, although the principle of inclusion-exclusion applies here, note that each multiple of 5^2 adds another power of 10, in addition to that already contributed by the multiple of 5. Hence, the terminating zeros are $\lfloor 123/5 \rfloor + \lfloor 123/25 \rfloor = 24 + 4 = 28$ in number.

Exercise 4.13. How many terminating zeros does 1234! have?

4.1.3 Permutations

Definition. By a *permutation* of elements we mean a particular ordering of the elements. For example, there are 6 different permutations of the three elements in $\{a, b, c\}$, i.e., abc, acb, bac, bca, cab, and cba.

Theorem 4.5. There are exactly $n!$ permutations of n distinct elements.

Proof. The first in the ordering can be any one of the n elements but, having chosen the first, there remain $n - 1$ choices for the second place, $n - 2$ for the third, etc. By the multiplication principle, there are $n(n-1)(n-2) \cdots 1 = n!$ permutations in all. ▽

Example. How many different permutations of the letters A, B, C, D, E, F which do not contain the string BAD in them?

Solution. There are $6! = 720$ permutations in all. However, a permutation such as $FEBADC$ must be counted out. Think of such arrangements as having only 4 elements, i.e., C, E, F, and BAD, which can be freely permuted as desired. Hence, there are $4! = 24$ forbidden permutations, leaving us with $720 - 24 = 696$ good ones.

Exercise 4.14. Count how many different permutations with the eight letters A, C, E, M, N, S, T, R, which have the following extra conditions.
a) Containing the string $CATS$.
b) Containing $CATS$ and MEN.
c) Containing either $CATS$ or MEN.
d) Containing neither $CATS$ nor MEN.

Question. How many permutations of the letters of the alphabet (from A to Z) can contain the string $COMPUTER$ and $BINARY$?

Test 4.15. How many permutations with the letters M, A, T, R, I, C, E, S contain either the string RAT or $MICE$, but not both?
a) 720
b) 828
c) 834
d) 840

Exercise* 4.16. In how many ways can the letters in M, A, T, R, I, C, E, S be rearranged, if no two vowels are allowed to be next to each other?

There are times we have to deal with permutations involving repeated items, such as with the collection of letters in the word $PEPPER$. We will use the term *multiset* to denote such a collection. Hence, a multiset is almost like a set, only that we allow repetition of its elements.

Theorem 4.6. The number of permutations on a multiset with n elements, k of which are identical, is given by $n!/k!$.

Proof. If the k repeated elements are labeled $a_1, a_2, \ldots, a_k$, there would be $n!$ distinct permutations. Without the labels, however, we overcount them as many times as $k!$, the number of permutations among themselves. Hence, $n!/k!$ gives the correct number of permutations on the multiset.   ▽

Example. Count the number of different rearrangements of the letters in the word $PEPPERED$.

Solution. If the three E's were labeled, the theorem would give us $8!/3!$ on account of the three identical P's. Without the labels, the E's can be permuted in $3!$ ways, each yielding the same arrangement. We conclude that the number of distinct permutations is $\frac{8!}{3!\,3!} = 1120$.

Test 4.17. Which one of these four words yields the most number of different rearrangements of its letters?
a) *UNUSUAL*
b) *EVERGREEN*
c) *REARRANGE*
d) *MISSISSIPPI*

Suppose that, in a class of 40 volunteers, three are to be selected as Chair, Treasurer, and Secretary, of a fund-raising project. In how many ways can the outcomes be determined? By the multiplication principle, assuming that a person is not allowed to take two positions, there are $40 \times 39 \times 38 = 40!/37!$ possibilities. This is a frequently encountered permutation principle which we state as a theorem next.

Definition. Let $P(n, k)$ denote the number of permutations of k elements, to be selected from a set with n elements, where $n \geq k$.

Theorem 4.7. As defined, $P(n, k) = \dfrac{n!}{(n - k)!}$.

Exercise 4.18. In a school's poetry contest, 15 poems have been submitted to the judges. In how many ways can the judges select first, second, and third winners, plus an honorable mention?

Suppose we wish to count the number of ways 5 people can be seated at a round table. This problem does involve permutations, but note that the table being round, an arrangement like *abcde* should not be distinguished from *bcdea*. This is an instance of the so-called *circular permutation,* which we discuss next.

Theorem 4.8. The number of circular permutations of k elements, chosen from a set with n elements, is given by $P(n, k)/k$. In particular, there are $(n - 1)!$ circular permutations with n elements.

Proof. The quantity $P(n, k)$ is too large for circular permutations, since each circular permutation like $a_1 a_2 \cdots a_k$ generates k different regular permutations, depending which element starts the cycle. In fact, $P(n, k)$ is k times larger than the number of circular permutations, and the result follows. $\triangledown$

Exercise 4.19. In a lab of 23 Biology students, five will be given the task of preparing the microscopes, one student per lab session on a rotating basis. In how many ways the rotating list of five students can be determined?

Exercise* 4.20. Six international leaders are to sit at a round table to discuss the World Peace Project. Amira and Elias are among them, and they sternly refuse to be seated next to each other, due to some unsettled personal issue. How many peaceful seating arrangements are possible?

Test 4.21. A necklace is to be constructed from 9 beads. There are 15 beads available, all distinguishable one from another. In how many ways can the necklace be made? Note that not only such a permutation is circular, but the necklace can also be turned over so that two arrangements like $a_1 a_2 \cdots a_9$ and $a_9 a_8 \cdots a_1$ are considered the same.

a) $P(15, 9)/2$
b) $P(15, 9)/9$
c) $P(15, 9)/18$
d) $P(15, 9)/81$

4.1.4   Combinations

The word *combination* can sometimes be used interchangeably with the word *set*, particularly when used in the context of a collection of elements. Hence, unlike permutations, a combination disregards ordering of elements.

Definition. Let $C(n, k)$ denote the number of different combinations of k elements, chosen out of a set with n elements, where $n \geq k$.

Theorem 4.9. As defined, $C(n, k) = \dfrac{n!}{k! \, (n - k)!}$. In particular, we have $C(n, k) = C(n, n - k)$.

Proof. If ordering were important, there would be $P(n, k)$ such selections. Without ordering, this quantity is an overcount by $k!$ times as many, since each selection can be permuted in that many ways without changing the combination. Hence, $C(n, k) = P(n, k)/k!$ as claimed. $\triangledown$

Note that the identity $C(n, k) = C(n, n - k)$ is consistent with the fact that one can choose k out of n elements by deselecting the unwanted $n - k$ elements. Note also that in computation, many terms appearing in the factorials will cancel out, e.g.,

$$C(23, 19) = \frac{23!}{19! \, 4!} = \frac{23 \times 22 \times 21 \times 20 \times 19!}{19! \times 4 \times 3 \times 2 \times 1} = 8855 \qquad (4.1)$$

Exercise* 4.22. Look at Equation (4.1) again carefully, and try to prove the following fact: The product of k consecutive integers is divisible by $k!$.

Example. If $|A| = 10$, how many subsets of A have at least 8 elements?

Solution. At least 8, in this case, means 8 or 9 or 10. By the addition principle, the answer we seek is given by $C(10, 8) + C(10, 9) + C(10, 10)$. Applying the theorem, we get $45 + 10 + 1 = 56$.

Question. What is the better way to count the number of subsets with at least 2 elements from a set with 20 elements?

Exercise 4.23. Let $|A| = 10$. Count how many subsets A has with the given additional conditions.
a) Only 6 or 7 elements.
b) At least 6 elements.
c) At most 4 elements.
d) At least 2 elements.

Exercise* 4.24. Prove that $C(n,0)+C(n,1)+C(n,2)+\cdots+C(n,n) = 2^n$.

Exercise 4.25. A typical domino card shows an unordered pair of numbers from 0 to 6. How many different cards does the game of dominoes involve?

Example. The Loyal Jordanian Airlines provides flights from Amman to 100 cities worldwide, at least three times a week for each destination. In the LJA weekly schedules, show that we can find two cities whose flights from Amman are on the exact same days of the week.

Solution. There are $\sum_{k\geq 3} C(7,k) = 99$ different combinations of three days or more of the week. By the pigeonhole principle, at least two destination cities will be assigned to the same set of flight days in any given week.

Exercise 4.26. At Planet University, 7469 students are preparing for the final exams, which will take place over five days. There are two exam sessions per day, one in the morning and another in the afternoon. Each student will have at least four exams to take, up to six for some. Show that there are 12 students, perhaps more, whose exam schedules are identical.

A combination problem may also involve a multiset. Consider, for instance, the number of ways one can purchase a dozen of donuts, where of course, choosing two or three of the same kind is not forbidden and neither is ordering the donuts necessary. We shall reveal how the next theorem applies in such situations.

Theorem 4.10. The number of non-negative integer solutions to the equation
$$x_1 + x_2 + \cdots + x_k = n$$
where both $k \geq 1$ and $n \geq 1$, is given by $C(n+k-1,n)$.

Proof. Consider a row of $n+k-1$ containers. There are $C(n+k-1,k-1)$ ways we can choose $k-1$ of these containers, in each of which we shall put the plus sign. Doing so partitions the row into k sections, with n empty containers in all. This gives a one-to-one correspondence between such selections and non-negative solutions to $x_1 + x_2 + \cdots + x_k = n$. (Two consecutive plus signs will squeeze in a zero container, hence a zero value of x_i.) The number of such solutions is then $C(n + k - 1, k - 1) = C(n + k - 1, n)$. ▽

Example. Back at Philadelphia Donuts, Elias is attracted to the "buy 6 get 6 free" offer. In how many ways can Elias choose 12 donuts, assuming that there are four kinds, each of which is plenty?

Solution. Elias is free to choose a non-negative number for each, as long as they sum to 12. Simply count the number of such solutions to $x_1 + x_2 + x_3 + x_4 = 12$. The formula gives $C(15, 12) = 910$.

Question. How does it affect the answer if Elias must have at least one of each kind of donut?

Example. Count the number of *positive* integer solutions to the equation $x + y + z = 12$.

Solution. To be able to apply the theorem, we introduce new variables: $X = x - 1$, $Y = y - 1$, and $Z = z - 1$. The number of positive solutions to $x + y + z = 12$ is the same as the number of non-negative solutions to $X + Y + Z = 9$, i.e., $C(11, 9) = 55$.

Exercise 4.27. Count the number of integer solutions to $x + y + z = 20$ with the given additional conditions.

a) All x, y, z positive
b) $x \geq 2$, $y \geq 3$, and $z \geq 4$.
c) $x \geq 3$ or $z \geq 5$.
d) $x \geq 3$, or $y \geq 4$, or $z \geq 5$.

Exercise* 4.28. How many four-digit integers have digit sum 10?

Test 4.29. Among quite a few men, women, boys, and girls at a refugee camp site, we are to select six representatives to meet the Queen. In how many ways the selection can be formed, if we require at least one woman or at least two girls to be included?

a) 20
b) 71
c) 91
d) 111

Exercise* 4.30. What is the number of non-negative integer solutions to the equation $x + y + z = 20$, given that $x \leq 10$, $y \leq 5$, and $z \leq 7$?

4.2 Introduction to Discrete Probability

By the probability of an event, we mean the likelihood that the event will occur, measured in a scale from 0 to 1. To make the idea more precise, we define the following terminologies.

Definition. When an experiment is conducted, e.g. tossing a pair of dice, the set of all possible outcomes is called the *sample space*. We assume, unless otherwise stated, that each outcome is equally likely. An *event* refers to any subset of the sample space S. When S is finite, we define the *probability* of an event E to be $p(E) = |E|/|S|$.

Example. A die is a small cube with six faces, labeled 1 to 6. What is the probability that when two dice are rolled, the numbers shown sum to 8?

Solution. There are 36 possible outcomes since we have a pair (a, b) of numbers, each with 6 possible values. The requested event E can occur with the pairs $(2, 6), (3, 5), (4, 4), (5, 3), (6, 2)$, and no other. Hence, $p(E) = 5/36$.

Exercise 4.31. Find the probability of each given event, with respect to rolling a pair of dice.

a) The numbers sum to 11.
b) A double, i.e., two equal numbers.
c) At least one number is a six.
d) The sum is at least seven.

Example. The letters a, b, c, d, e are randomly rearranged. What is the probability that the two vowels are next to each other?

Solution. The sample space has 5! outcomes, of which 4! contain the string ae, and another 4! for ea. The probability that a is next to e is then $2 \times 4!/5! = 0.4$.

Exercise 4.32. In a school's poetry contest, 15 poems have been submitted to the judges and there will be a first, second, and third winners, plus an honorable mention. Amira, who is not very talented at writing poems, has submitted hers too. Assuming the best scenario (for Amira) in which the judges randomly choose the winners, what is the probability that Amira will get an award for her poem?

Test 4.33. Suppose that three dice are rolled. Which one of these four events is the least likely to occur?

a) Three distinct numbers.
b) The sum is at most five.

c) At least two sixes.
d) No odd numbers appear.

The following set properties are quite easy to obtain.

Theorem 4.11. Let E and F be events, subsets of the sample space S.

1) If $E \subseteq F$, then $p(E) \leq p(F)$. In particular, $p(\emptyset) = 0 \leq p(E) \leq 1 = p(S)$.

2) $p(\neg E) = 1 - p(E)$

3) $p(E \cup F) = p(E) + p(F) - p(E \cap F)$

4) $p(E - F) = p(E) - p(E \cap F)$

Example. A number is randomly selected between 1 and 1000 inclusive. What is the probability that the number is a multiple of 15 but not of 9?

Solution. By Theorem 4.11(4), the probability is

$$\frac{\lfloor 1000/15 \rfloor}{1000} - \frac{\lfloor 1000/45 \rfloor}{1000} = \frac{66 - 22}{1000} = 0.044$$

where $45 = \operatorname{lcm}(15, 9)$.

Exercise 4.34. A coin is tossed 10 times in succession. What is the probability of each event given below?

a) Equal number of heads and tails.
b) Heads turn up the first three tosses or the last three tosses.
c) Tails turn up no more than eight times.
d) Exactly four heads but no two of them are consecutive.

Exercise* 4.35. Five men and five women are having dinner at Tea Kitchen Chinese restaurant, where their reserved table is circular. What is the probability that the ten people are seated such that men and women alternate?

Definition. The probability of an event E, given that event F has already occurred, or is assumed to occur, is defined to be

$$p(E|F) = \frac{p(E \cap F)}{p(F)}$$

The notation $p(E|F)$ is read *the probability of E given F*, and such is referred to as a *conditional probability*. Since we assume that all possible outcomes are equally likely, note that $p(E|F) = |E \cap F|/|F|$.

Example. A coin is tossed 5 times in a row. What is the probability of getting at least two heads, given that a tail turns up in the first toss?

Solution. Since the first toss is already determined, there are $2^4 = 16$ possible outcomes for the event F. Of these, one outcome consists of all tails, and four with exactly one head. Hence, the event E of at least two heads has $|E| = 16 - 5 = 11$. We conclude that $p(E|F) = 11/16$.

Exercise 4.36. Repeat Exercise 4.34, this time under the given condition that the last two tosses turn up tails.

Definition. Two events, E and F, are said to be *independent* of each other when

$$p(E \cap F) = p(E) \times p(F)$$

This says that the likelihood of E occurring has no effect on that of F, vice versa. Indeed, this definition implies that E and F are independent events if and only if $p(E|F) = p(E)$ and $p(F|E) = p(F)$.

Example. A newly wed couple plans to have three children, no more no less. What is the probability that the first two children are girls and the third a boy?

Solution. It is fair to assume that the gender of each child is not affected by that of the siblings. Hence, the probability here is $1/2 \times 1/2 \times 1/2 = 1/8$. Note that the result agrees with the fact that the sample space contains eight elements, i.e., *BBB, BBG, BGB, BGG, GBB, GBG, GGB, GGG.*

Exercise 4.37. Repeat Exercise 4.34(b), assuming independent events.

Exercise* 4.38. What is the probability that, in a group of 23 strangers, two share the same birthday? Hint: it is more likely than not. In fact, this famous problem is named the *birthday paradox*. You may ignore leap years.

Appendix: Playing Cards

A deck of playing cards consists of 52 cards, of 13 different kinds each: 2, 3, 4, 5, 6, 7, 8 ,9, 10, Jack, Queen, King, and Ace. Each kind comes in four different suits, whose names and colors are listed in Table 4.1

Table 4.1: The four suits of playing cards.

Symbol	Name	Color
♠	spade	black
♡	heart	red
♢	diamond	red
♣	club	black

Exercise 4.39. Suppose that a hand of four cards is dealt from the deck of 52 cards. Find the probability of each event given below.

a) All four cards are red.
b) Exactly two queens and at least one king.
c) At least two nines or at least one heart.
d) At least three different suits.

Test 4.40. Three cards are drawn from a deck of 52. Which one of the following four events is the most likely to occur?

a) All three are of the same suit.
b) At least two are of a kind.
c) All three are of the same color.
d) At least one is a king or queen.

Example. In a game of poker, a hand of five cards is dealt. Each kind, from 2 to Ace, corresponds to a numerical value from 2 to 14, respectively. As an exception, the ace can alternately be a 1, instead of 14, to the player's advantage depending on the particular hand. Note that there are $C(52,5) = 2,598,960$ different poker hands. Order the following events according to their probabilities.

a) Four-of-a-kind: four cards are of the same kind.
b) Flush: all five are of the same suit.
c) Royal flush: an ace, a king, a queen, a jack and a 10, all of the same suit.
d) Straight: the five cards have consecutive numerical values.

Solution. It suffices to consider the cardinality of each event.

a) There are 13 kinds to consider. Since there are exactly four cards of each kind, we are left to choose a fifth card among the remaining 48. By the multiplication principle, there are $13 \times 48 = 624$ four-of-a-kind hands.
b) There are four suits to pick, in each we choose five from 13. Thus, $4 \times C(13,5) = 5148$ flush hands.
c) There are only four royal flushes, of course.
d) The least value in a straight hand is between 1 and 10. For each card in such hand, there are four suits to pick. This gives $10 \times 4^5 = 10240$ in all.

Exercise 4.41. Arrange four more poker events described below, together with the four from the preceding example, in the order of their probability. The resulting list, from the least probability to the greatest, determines their order of superiority in the poker game.

a) Straight flush: flush and straight simultaneously.
b) Full house: three cards of a kind, two of another kind.
c) Three-of-a-kind: three cards are of the same kind; the other two are of different kinds.

d) Two pairs: two of a kind, another pair of a second kind, and a fifth card of a third kind.

Exercise* 4.42. A deck of playing cards can sometimes include two jokers, one colored and another black and white, making a total of 54 cards. The jokers are wild cards, in the sense that a joker can substitute for any other card as the player wishes. For instance, with three kings, a two, and a joker, a player may decide to treat the joker as another king, making the hand a four-of-a-kind. If a 54-card deck is used, analyze how this would affect the order of superiority among the eight poker events discussed earlier. Interestingly, note that it is now possible to have a five-of-a-kind!

4.3 The Binomial Coefficients

A *binomial coefficient*, denoted by $\binom{n}{k}$, counts the number of ways one can choose k objects, ordering ignored, from a set of n elements. The notation $\binom{n}{k}$ is often read n *choose* k, and is none other than the quantity $C(n, k)$. Hence,

$$\binom{n}{k} = \frac{n!}{k! \, (n - k)!} \tag{4.2}$$

and in particular, we also have $\binom{n}{1} = n$, $\binom{n}{n} = 1$, and $\binom{n}{k} = \binom{n}{n-k}$.

Binomial coefficients get their name from the fact that they appear as the coefficients in the expansion of the binomial $(x + y)^n$ over the real numbers. This is the familiar binomial theorem.

Theorem 4.12 (The Binomial Theorem). For $n \geq 1$,

$$(x + y)^n = \sum_{k=0}^{n} \binom{n}{k} x^{n-k} y^k$$

Proof. We write $(x + y)^n = (x + y)(x + y) \cdots (x + y)$, with n factors of $(x + y)$. When multiplying this out, each of the n factors will contribute one exponent of x or of y, yielding in the end a term of the form $x^{n-k} y^k$. How many like terms do we have for each such form? The $n - k$ exponents of x can come from choosing any $n - k$ out of the n factors $(x + y)$, while by default, the remaining k goes to y. This is exactly the quantity $\binom{n}{n-k} = \binom{n}{k}$, which becomes the coefficient of $x^{n-k} y^k$ for each $0 \leq k \leq n$. $\triangledown$

Question. How can one reestablish the identity $\sum \binom{n}{k} = 2^n$ using the binomial theorem?

Exercise 4.43. Prove the identity $k\binom{n}{k} = n\binom{n-1}{k-1}$, where $n \geq k \geq 1$, and use it to establish another identity: $\sum k\binom{n}{k} = n2^{n-1}$.

The proof of Theorem 4.12 uses a *combinatorial argument*. Many identities involving the binomial coefficients can be proved in two ways—the other way using the explicit formula given in (4.2).

Exercise* 4.44. Prove the binomial theorem one more time, this time using mathematical induction and the formula given in (4.2).

The famous *Pascal's triangle* is a matrical arrangement of the binomial coefficients, where $\binom{n}{k}$ is the (n, k) entry in the array. Since $k \leq n$, the "matrix" would be lower triangular. To make the triangle more presentable, we center justify the rows. The first five rows of Pascal's triangle is depicted below.

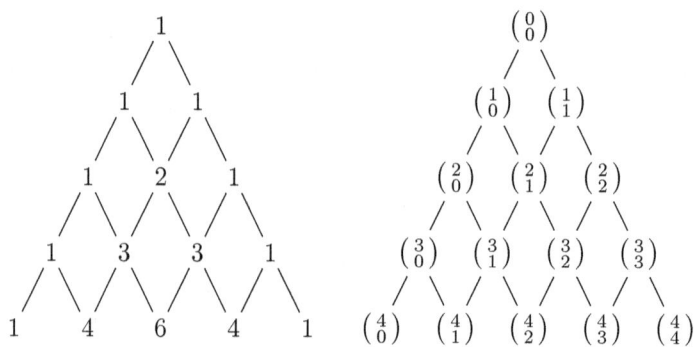

The grids are added to show a particular relation in which two consecutive coefficients in a given row sum to the binomial coefficient directly below them. We shall now prove this Pascal's formula.

Theorem 4.13 (Pascal's Formula). For $n \geq k \geq 0$,

$$\binom{n}{k} + \binom{n}{k+1} = \binom{n+1}{k+1}$$

Proof. Using the formula for $\binom{n}{k}$ given in (4.2),

$$\binom{n}{k} + \binom{n}{k+1} = \frac{n!}{k!\,(n-k)!} + \frac{n!}{(k+1)!\,(n-k-1)!}$$

$$= \frac{(k+1)\,n! + (n-k)\,n!}{(k+1)!\,(n-k)!} = \frac{(n+1)\,n!}{(k+1)!\,(n-k)!}$$

The last expression is indeed the formula for $\binom{n+1}{k+1}$.   ▽

Exercise 4.45. With the help of Pascal's triangle, expand the binomial $(x+y)^8$.

Exercise* 4.46. For the third time again, prove the identity $\sum\binom{n}{k} = 2^n$, this time by induction, employing Pascal's formula at the induction step.

As a practice, we shall illustrate how Pascal's formula can also be derived using a combinatorial argument.

Example. Prove Pascal's formula again using a combinatorial argument.

Solution. From a group of $n+1$ students, $k+1$ are to be selected to partic-ipate in a school's play. There are then $\binom{n+1}{k+1}$ ways to do this. Elias is one of these students. Now the selection can either include Elias, in which case k more students will be chosen among the remaining n, or can be without Elias, in which case all $k+1$ will be chosen from the remaining number. By the addition principle, we get $\binom{n+1}{k+1} = \binom{n}{k} + \binom{n}{k+1}$.

Exercise 4.47. Prove the following claims in the given sequence.
a) Show that $\sum(-1)^k\binom{n}{k} = 0$ using the binomial theorem.
b) Conclude that $\sum_{k \text{ odd}}\binom{n}{k} = \sum_{k \text{ even}}\binom{n}{k}$.
c) There are 2^{n-1} ways to choose an even number of elements out of n elements, and the same number of ways for odd.
d) Find a combinatorial argument for (c).

Exercise 4.48. Prove the identity $\sum\binom{n}{k}^2 = \binom{2n}{n}$ using a combinatorial argument, by considering a group of n men and n women, from which n persons shall be selected.

To conclude this section we shall reproduce Fermat's little theorem (The-orem 1.12) but this time with a proof by induction, employing only the properties of binomial coefficients. We need first the following theorem.

Theorem 4.14. If p is a prime, then $\binom{p}{k}$ is a multiple of p, for as long as $1 \le k \le p-1$.

Proof. The formula

$$\binom{p}{k} = \frac{p!}{k!\,(p-k)!}$$

shows that the numerator is a multiple of p, while the denominator will factor into primes, all less than p and none of which can cancel out the p from the numerator. Hence, the quantity $\binom{p}{k}$ is a multiple of p. ▽

Theorem 4.15 (Fermat's Little Theorem). Let p be a prime number and a be an integer in the range $0 \le a < p$. Then $a^p \bmod p = a$.

Proof. The claim is obvious for $a = 0$. By the binomial theorem,

$$(a+1)^p = a^p + \binom{p}{1} a^{p-1} + \binom{p}{2} a^{p-2} + \cdots + \binom{p}{p-1} a + 1$$

Theorem 4.14 says that all these terms, except the first and the last, are multiples of p. It follows that $(a+1)^p \equiv a^p + 1 \pmod{p}$. (See Section 1.2.1 for congruence.) So if $a^p \bmod p = a$ holds for some $a \le p - 2$, then $(a+1)^p \bmod p = a + 1$, and the proof completes by induction.   ▽

Question. Why does this Fermat's little theorem look different from Theorem 1.12 or 3.30?

Exercise* 4.49. Use induction on k, with Theorem 4.14 and Pascal's formula again, to show that for any prime $p > 2$ and for $0 \le k \le p - 1$, the congruence $\binom{p-1}{k} \equiv (-1)^k \pmod{p}$ holds.

4.4   Recursive Sequences

By a *sequence* S_n we mean a function $S(n)$ whose domain is the integers, usually non-negative; hence, interchangeably, $S(n)$ also refers to the enumerated range values, $S(0), S(1), S(2), S(3), \ldots$ For instance, with $S(n) = n^2$, the sequence S_n consists of $0, 1, 4, 9, 16, 25, \ldots$

Exercise 4.50. Find a suitable function $S(n)$ for each sequence below.
a) $1, 2, 4, 8, 16, 32, 64, 128, \ldots$
b) $4, 5, 7, 11, 19, 35, 67, 131, \ldots$
c) $7, 11, 15, 19, 23, 27, 31, 35, \ldots$
d) $5, 6, 9, 14, 21, 30, 41, 54, \ldots$

Test 4.51. Which function generates a sequence of zeros and ones?
a) $S_n = \lfloor \frac{n}{2} \rfloor$
b) $S_n = \lfloor \frac{n}{n+1} \rfloor$
c) $S_n = n - 2 \lfloor \frac{n}{2} \rfloor$
d) $S_n = n - \lfloor \sqrt{n} \rfloor^2$

Definition. A sequence S_n is called *recursive* if the function $S(n)$ is given in terms of $S(0), S(1), \ldots, S(n-1)$.

For example, the function $S(n) = 2S(n-1)$, assuming initial value $S(0) = 1$, generates the recursive sequence $1, 2, 4, 8, 16, 32, 64, \ldots$ Of course, this is none other than the sequence $S_n = 2^n$.

In counting theory, we may encounter sequences which are easier to represent recursively, rather than explicitly. And in some cases, the recursively obtained expression may then lead to finding the explicit function $S(n)$ or its particular value at a given n.

Example. Let S_n count the number of series of sum n consisting of ones and twos. For example, $S_4 = 5$, since there are five such series: $2 + 2 = 2 + 1 + 1 = 1 + 2 + 1 = 1 + 1 + 2 = 1 + 1 + 1 + 1$. Express S_n as a recursive sequence.

Solution. Such a series can be obtained in two ways: by adding 1 to a similar series whose sum is $n - 1$, or by adding 2 to one with sum $n - 2$. Hence, we obtain $S_n = S_{n-1} + S_{n-2}$. The initial conditions $S_1 = 1$ and $S_2 = 2$ define the unique sequence S_n, whose first few terms are

$$1, 2, 3, 5, 8, 13, 21, 34, 55, 89, 144, \ldots$$

In fact this is the famous Fibonacci sequence, which we shall discuss at length after learning next how to deal with such recursive sequences.

Exercise 4.52. Let S_n count the number of series of sum n, using only the numbers 1, 2, 3. Find a recursive expression for the function $S(n)$, and use it to compute S_{10}.

4.4.1 Solving Linear Recurrence

We refer to the recursive expression of S_n by calling it the *recurrence relation* of the sequence. In general, finding an explicit formula for S_n is not an easy task. We shall discuss only *linear homogeneous* relations, i.e., of the form

$$S_n = a_1 S_{n-1} + a_2 S_{n-2} + \cdots + a_k S_{n-k}$$

with constant coefficients. In particular, with the *degree* $k = 2$, the solution for S_n we will present in two separate cases.

Exercise* 4.53. The nth *Fermat number* is defined by $F_n = 2^{2^n} + 1$, for $n \geq 0$. Show that F_n satisfies the quadratic non-homogeneous recurrence relation of degree one, $F_n = F_{n-1}^2 - 2F_{n-1} + 2$.

Theorem 4.16. Let a and $b \in \mathbb{R}$, and let x_1 and x_2 be the two roots of $x^2 = ax + b$. If $x_1 \neq x_2$, then the recurrence relation $S_n = aS_{n-1} + bS_{n-2}$ holds if and only if $S_n = cx_1^n + dx_2^n$, for some real numbers c and d.

Proof. We show first that $S_n = cx_1^n + dx_2^n$ obeys the recurrence relation:

$$
\begin{aligned}
aS_{n-1} + bS_{n-2} &= a(cx_1^{n-1} + dx_2^{n-1}) + b(cx_1^{n-2} + dx_2^{n-2}) \\
&= cx_1^{n-2}(ax_1 + b) + dx_2^{n-2}(ax_2 + b) \\
&= cx_1^{n-2}x_1^2 + dx_2^{n-2}x_2^2 \\
&= cx_1^n + dx_2^n \\
&= S_n
\end{aligned}
$$

Conversely, suppose that a sequence is given by the recurrence relation $S_n = aS_{n-1} + bS_{n-2}$. Note that the first two initial values, S_0 and S_1, completely determine the sequence generated by S_n. Since $x_1 \neq x_2$, we may find two constants, c and d, defined by

$$c = \frac{S_1 - S_0 x_2}{x_1 - x_2} \quad \text{and} \quad d = S_0 - c \tag{4.3}$$

Now let $T_n = cx_1^n + dx_2^n$, and observe that the sequence T_n satisfies the same initial conditions as does S_n:

$$T_0 = c + d = S_0$$
$$T_1 = cx_1 + dx_2 = cx_1 + (S_0 - c)x_2 = S_0 x_2 + c(x_1 - x_2) = S_1$$

And, as before, $T_n = aT_{n-1} + bT_{n-2}$. It follows that $T_n = S_n$ for all $n \geq 0$, and the proof is complete. ▽

Note that Theorem 4.16, with its proof, contains an algorithm for finding the explicit formula corresponding to the recurrence relation, provided that S_0 and S_1 have been fixed.

Example. Solve the recurrence relation $S_n = S_{n-1} + 2S_{n-2}$, with initial conditions $S_0 = 1$ and $S_1 = 2$.

Solution. We have to find the roots of $x^2 = x + 2$, i.e., $(x+1)(x-2) = 0$. They are distinct: $x_1 = -1$ and $x_2 = 2$. Our solution will be in the form $S_n = c(-1)^n + d(2^n)$. To find c and d, we call for S_0 and S_1:

$$S_0 = 1 = c + d \quad \text{and} \quad S_1 = 2 = -c + 2d$$

Adding the two equations yields $3 = 3d$. Hence, $d = 1$ and $c = 0$. (Or we could use (4.3) to solve for c and d.) The explicit formula turns out to be the familiar $S_n = 2^n$, which agrees with the fact that the relation $S_n = S_{n-1} + 2S_{n-2}$ does generate the sequence $1, 2, 4, 8, 16, 32, 64, \ldots$, i.e., the powers of 2.

Question. In the preceding example, what would be the better way to *prove* that $S_n = 2^n$ without solving the recurrence?

Example. Find an explicit formula for the number of series with ones and twos, whose sum equals n.

Solution. We have seen that $S_n = S_{n-1} + S_{n-2}$ with initial values $S_1 = 1$ and $S_2 = 2$. It does no harm if we define $S_0 = 1$. Using the quadratic formula, we find that $x^2 = x + 1$ has two roots given by $x = \frac{1}{2}(1 \pm \sqrt{5})$. Hence, $2^n S_n = c(1 + \sqrt{5})^n + d(1 - \sqrt{5})^n$, and

$$S_0 = 1 = c + d$$
$$2S_1 = 2 = c(1 + \sqrt{5}) + d(1 - \sqrt{5})$$

Omitting algebraic details, we solve for c and d by substitution and arrive at the not-so-attractive formula

$$S_n = \left(\frac{5 + \sqrt{5}}{10}\right)\left(\frac{1 + \sqrt{5}}{2}\right)^n + \left(\frac{5 - \sqrt{5}}{10}\right)\left(\frac{1 - \sqrt{5}}{2}\right)^n$$

Exercise 4.54. Find an explicit formula for each given recurrence relation.
a) $S_n = S_{n-1} + 6S_{n-2}$; $S_0 = 3$; $S_1 = 4$
b) $S_n = 3S_{n-1} - 2S_{n-2}$; $S_0 = 1$; $S_1 = 3$
c) $S_n = S_{n-1} + 3S_{n-2}$; $S_0 = 1$; $S_1 = 1$
d) $S_n = 6S_{n-1} - 8S_{n-2}$; $S_0 = 1$; $S_1 = 5$

Exercise* 4.55. A rectangular floor of size $2 \times n$ is to be tiled using 1×2 and 2×2 pieces. Find a recurrence relation for S_n, the number of ways the tiling can be done, and then find an explicit formula.

Before proceeding to the case $x_1 = x_2$, we may state that Theorem 4.16 generalizes in a rather orderly manner to recurrence relations of higher degree, as follows.

Theorem 4.17. Let $x_1, x_2, \ldots, x_k$ be distinct roots to the equation

$$x^k = a_1 x^{k-1} + a_2 x^{k-2} + \cdots + a_{k-1}x + a_k \tag{4.4}$$

Then the recurrence relation $S_n = a_1 S_{n-1} + a_2 S_{n-2} + \cdots + a_k S_{n-k}$ holds if and only if

$$S_n = c_1 x_1^n + c_2 x_2^n + \cdots + c_k x_k^n \tag{4.5}$$

with some constants c_i.

To prove this claim, one can show that (4.5) obeys the given recurrence relation. Then, as with the case $k = 2$, we are left to establishing the fact that the system of k equations, given by (4.5) for $0 \leq n \leq k - 1$, has a unique solution $(c_1, c_2, \ldots, c_k)$. This follows from linear algebra, where the matrical equation $[S] = [X][C]$ is determined by $[C] = [X]^{-1}[S]$. In fact,

$$[X] = \begin{bmatrix} 1 & 1 & \cdots & 1 \\ x_1 & x_2 & \cdots & x_k \\ x_1^2 & x_2^2 & \cdots & x_k^2 \\ \cdots & \cdots & \cdots & \cdots \\ x_1^{k-1} & x_2^{k-1} & \cdots & x_k^{k-1} \end{bmatrix}$$

is a *Vandermonde matrix* of non-zero determinant $\det[X] = \prod_{i \geq j}(x_i - x_j)$.

Exercise 4.56. Solve the recurrence relation $S_n = 2S_{n-1} + S_{n-2} - 2S_{n-3}$ subject to the initial values $S_0 = 0$, $S_1 = 5$, $S_2 = 3$.

Theorem 4.18. Let $x = r$ be the unique root of $x^2 = ax + b$ over the real numbers. Then the recurrence relation $S_n = aS_{n-1} + bS_{n-2}$ holds if and only if $S_n = (cn + S_0)r^n$ for some constant c.

Proof. Before we show that $S_n = (cn + S_0)r^n$ meets the given recurrence, we need the fact that $x^2 - ax - b = (x - r)^2$, so that $a = 2r$ and $b = -r^2$. In particular, $ar + 2b = 0$. Now,

$$
\begin{aligned}
aS_{n-1} + bS_{n-2} &= a\big(c(n-1)r^{n-1} + S_0 r^{n-1}\big) + b\big(c(n-2)r^{n-2} + S_0 r^{n-2}\big) \\
&= cr^{n-2}\big(a(n-1)r + b(n-2)\big) + S_0 r^{n-2}(ar + b) \\
&= cr^{n-2}\big(n(ar + b) - (ar + 2b)\big) + S_0 r^{n-2}(ar + b) \\
&= cr^{n-2}n(ar + b) + S_0 r^{n-2}(ar + b) \\
&= cnr^n + S_0 r^n = S_n
\end{aligned}
$$

where we have substituted $ar + b = r^2$.

Lastly, if $T_n = (cn + S_0)r^n$, with the choice of $c = S_1/r - S_0$, then we can show that $T_0 = S_0$ and $T_1 = S_1$, so that $T_n = S_n$ for all $n \geq 0$.   ▽

Example. Solve the recurrence relation $S_n = 4S_{n-1} - 4S_{n-2}$ under the initial conditions $S_0 = 0$ and $S_1 = 1$.

Solution. We find that $x^2 - 4x + 4 = 0$ has a unique root $x = 2$, hence $S_n = (cn + 0)2^n$. Substituting $S_1 = 1$, we get $c = 1/2$. Thus, the function $S_n = n \times 2^{n-1}$.

Exercise 4.57. Solve the recurrence relation $S_n = 6S_{n-1} - 9S_{n-2}$ with the initial values $S_0 = 2$ and $S_1 = 3$.

Exercise* 4.58. Let $S_n = 4S_{n-1} - 4S_{n-2}$ with $S_0 = 1$ and $S_1 = 2$. Use mathematical induction to prove that $S_n = 2^n$ for all $n \geq 0$.

The polynomial equation appearing in (4.4) is called the *characteristic equation* of the recurrence relation $S_n = a_1 S_{n-1} + a_2 S_{n-2} + \cdots + a_k S_{n-k}$. To complete our discussion, we merely state that in general we will have the solution $S_n = \sum S_i$, where each term $S_i = S$ is given by

$$
S = (c_{m-1}n^{m-1} + \cdots + c_2 n^2 + c_1 n + c_0)r^n
$$

for every root r of multiplicity m of the characteristic equation.

Example. Suppose that $S_n = 5S_{n-1} - 6S_{n-2} - 4S_{n-3} + 8S_{n-4}$. Then the characteristic equation $x^4 - 5x^3 + 6x^2 + 4x - 8 = (x-2)^3(x+1)$ reveals that S_n is given explicitly by $S_n = (c_2n^2 + c_1n + c_0)2^n + d(-1)^n$ for some constants c_i and d. These constants are uniquely determined by the initial conditions, in this case, S_0, S_1, S_2, and S_3.

Exercise 4.59. For each given characteristic equation, write the corresponding recurrence relation for S_n and its explicit function.

a) $0 = (x-2)^4$
b) $0 = (x-3)^2(x+6)$
c) $0 = (x-1)^3(x+1)^3$
d) $0 = (x-2)^2(x-3)(x+3)$

4.4.2 Generating Functions

Definition. The *generating function* for a given sequence S_n is the power series $S(x)$ given by

$$S(x) = \sum_{n=0}^{\infty} S_n x^n$$

For example, the sequence $S_n = 1$ gives rise to the generating function $S(x) = 1 + x + x^2 + x^3 + \cdots = \frac{1}{1-x}$. Another example, the polynomial $S(x) = (1+x)^m$ can be treated as the generating function for the sequence $S_n = \binom{m}{n}$ for $n \leq m$ and $S_n = 0$ for $n > m$.

Exercise 4.60. Find the generating function for each given sequence, and use your knowledge of Taylor series to write it in a compact form.

a) $S_n = (-1)^n$
b) $S_n = 1/n!$
c) $S_n = (-2)^n/n!$
d) $S_n = n \bmod 2$

Exercise* 4.61. Recall the Euler's phi function, defined by $\phi(n) = |\mathbb{U}_n|$ in Theorem 3.30. Find the generating function for the sequence $\phi(p^n)$, where p is a prime number.

We are not concerned with the interval of convergence of the power series, but we shall illustrate with a few examples of how a generating function can sometimes help in solving a combinatorial problem or a recurrence relation.

Theorem 4.19. For a fixed $k \geq 1$, we have

$$\sum_{n=0}^{\infty} \binom{n+k-1}{n} x^n = \left(\frac{1}{1-x}\right)^k$$

Proof. Consider the function $S(x) = (1+x+x^2+x^3+\cdots)^k$. The coefficient of the x^n-term in this power series is determined by the number of k non-negative exponents which sum to n. Hence, according to Theorem 4.10, $S(x)$ is the generating function for the sequence $S_n = C(n+k-1, n)$. ▽

Exercise 4.62. Derive the generating function for the sequence $S_n = n^2$ by letting $k = 2$ in Theorem 4.19 and differentiating the power series at some point.

Exercise* 4.63. Let p_n denote the number of partitions of n using positive integers. For example, $p_5 = 7$ because we have $5 = 4+1 = 3+2 = 3+1+1 = 2+2+1 = 2+1+1+1 = 1+1+1+1+1$. Show why the generating function for p_n is given by

$$\sum_{n=0}^{\infty} p_n x^n = \prod_{n=1}^{\infty} \frac{1}{1-x^n}$$

Example. Amira is to buy a box of n donuts, desiring an even number of blueberry, at most one lemon, at most two sugar, and an arbitrary number of cherry. There are also plain donuts, which she wants a multiple of three. In term of n, how many different combinations of donuts can Amira buy?

Solution. Let S_n represent this quantity. In the order given above, the generating function for S_n can be written as the following product.

$$S(x) = (1 + x^2 + x^4 + \cdots) \times (1 + x) \times (1 + x + x^2) \times$$
$$(1 + x + x^2 + \cdots) \times (1 + x^3 + x^6 + \cdots)$$
$$= \left(\frac{1}{1-x^2}\right) \left(\frac{1-x^2}{1-x}\right) \left(\frac{1-x^3}{1-x}\right) \left(\frac{1}{1-x}\right) \left(\frac{1}{1-x^3}\right)$$
$$= \frac{1}{(1-x)^3}$$

Theorem 4.19 applies to give us $S_n = \binom{n+2}{n} = \frac{1}{2}(n+2)(n+1)$.

It is evidently deliberate that, in the preceding example, the five terms cancel out nicely. In general we may not be that lucky, but generating functions can nevertheless be ready to obtain where an explicit expression for S_n is not.

Exercise 4.64. Find the generating function for S_n, the number of non-negative integer solutions to $x_1 + x_2 + x_3 + x_4 = n$, subject to the given additional conditions.

a) Each x_i is odd.

b) $x_i \geq i$ for each.
c) $x_1 \leq 4$ and x_4 is a multiple of 5.
d) x_1 is odd, $x_2 \geq 2$, x_3 is even, and $x_4 \leq 9$.

Exercise 4.65. Find the generating function for the number of combinations of pennies, nickels, dimes, and quarters, which in all make n cents.

Exercise* 4.66. Find the number of non-negative integer solutions to $x_1 + x_2 + x_3 + x_4 + x_5 = n$ given that $x_3 \leq 2$, while x_5 is twice larger than x_1.

Example. Let us solve the recurrence relation $S_n = 3S_{n-1} + 4S_{n-2}$, this time employing the generating function for S_n. Assume $S_0 = 1$ and $S_1 = 2$.

Solution. Let $S(x) = \sum S_n x^n$. Then,

$$
\begin{aligned}
S(x) = \quad & S_0 && + S_1 x && + S_2 x^2 && + S_3 x^3 && + \quad \cdots \\
-3x S(x) = \quad & && - 3S_0 x && - 3S_1 x^2 && - 3S_2 x^3 && - \quad \cdots \\
-4x^2 S(x) = \quad & && && - 4S_0 x^2 && - 4S_1 x^3 && - \quad \cdots
\end{aligned}
$$

Due to the relation $S_n = 3S_{n-1} + 4S_{n-2}$, from the second exponent onward, the like terms vertically sum to zero. Hence,

$$(1 - 3x - 4x^2)S(x) = S_0 + S_1 x - 3S_0 x = 1 - x$$

from which we obtain

$$S(x) = \frac{1-x}{1 - 3x - 4x^2} = \frac{1-x}{(1+x)(1-4x)} = \frac{c}{1+x} + \frac{d}{1-4x}$$

The last expression, involving constants c and d, is the method of partial fractions we borrow from Calculus. To find these constants, we equate the numerators:

$$1 - x = c(1 - 4x) + d(1 + x)$$

Comparing like terms,

$$1 = c + d \quad \text{and} \quad -1 = -4c + d$$

which has simultaneous solution $c = 2/5$ and $d = 3/5$. Hence,

$$
\begin{aligned}
S(x) &= \frac{2/5}{1+x} + \frac{3/5}{1-4x} \\
&= \frac{2}{5}(1 - x + x^2 - x^3 + \cdots) + \frac{3}{5}(1 + 4x + 16x^2 + 64x^3 + \cdots)
\end{aligned}
$$

The coefficient of x^n in $S(x)$ is therefore $S_n = \frac{2}{5}(-1)^n + \frac{3}{5}(4^n)$.

Exercise 4.67. Redo Exercise 4.57 using generating function. Recall the method of partial fractions for multiple roots, e.g.,

$$\frac{1}{(ax+b)^2} = \frac{c}{ax+b} + \frac{d}{(ax+b)^2}$$

Exercise 4.68. Solve again Exercise 4.56, by generating function.

So far, generating functions seem to work suitably in our counting problems, all of which involve combinations. Where permutations are concerned, however, it is best to approach the problem with a slightly modified generating function.

Definition. The *exponential generating function* for a sequence S_n is the power series $S^*(x)$ given by

$$S^*(x) = \sum_{n=0}^{\infty} \frac{S_n}{n!} x^n$$

For example, the sequence $S_n = 1$ is now represented by the exponential generating function $S^*(x) = 1 + \frac{x}{1!} + \frac{x^2}{2!} + \frac{x^3}{3!} + \cdots = e^x$. Similarly, the polynomial $S^*(x) = (1+x)^m$ is the exponential generating function for the sequence $S_n = \binom{m}{n}/n! = P(m,n)$ for $n \le m$.

Exercise 4.69. Find the exponential generating function for each given sequence, written in a compact form.
a) $S_n = (-1)^n$
b) $S_n = 2^{-n}$
c) $S_n = n$
d) $S_n = n \bmod 2$

Theorem 4.20. For a fixed $k \ge 1$, we have

$$\sum_{n=0}^{\infty} \frac{k^n}{n!} x^n = \left(1 + \frac{x}{1!} + \frac{x^2}{2!} + \frac{x^3}{3!} + \cdots \right)^k$$

Proof. It is clear that both sides equal e^{kx}. Nevertheless, we shall give a useful combinatorial argument as follows. Let $S_n = k^n$, the number of permutations of n elements, chosen from k different objects, allowing repetitions. Now the "coefficient" of $\frac{x^n}{n!}$ upon expanding the k products on the right is of the form

$$\sum_{m_1+\cdots+m_k=n} \frac{n!}{m_1! \times m_2! \times \cdots \times m_k!}$$

Each term in the summation is precisely the number of permutations of n elements consisting of m_i copies of the ith object, based on Theorem 4.6. We conclude that the resulting exponential generating function is indeed the one for $S_n = k^n$. ▽

Example. Determine the formula S_n for the number of n-digit integers consisting of only the digits 1 to 5, and where the digits 2 and 4 must each appear an odd number of times.

Solution. The exponential generating function for S_n would be the product of five, each corresponding to the digits 1 to 5, respectively:

$$S^*(x) = e^x \times \left(\frac{x}{1!} + \frac{x^3}{3!} + \frac{x^5}{5!} + \cdots \right) \times e^x \times \left(\frac{x}{1!} + \frac{x^3}{3!} + \frac{x^5}{5!} + \cdots \right) \times e^x$$

$$= e^{3x} \left(\frac{e^x - e^{-x}}{2} \right)^2 = \frac{1}{4}(e^{5x} - 2e^{3x} + e^x)$$

Expanding the series again, we see that $S_n = \frac{1}{4}(5^n - 2 \times 3^n + 1)$.

Exercise 4.70. Use exponential generating functions to count the number of n-letter words which can be formed using the five vowels a, e, i, o, u, allowing repetitions, and assuming the given additional conditions.

a) Each one except a and e must appear an even number of times.
b) The number of a's is even, while e's odd.
c) The vowel a is used at least once, e at least twice.
d) Both a and e are not allowed to repeat.

4.4.3 Fibonacci Numbers

We conclude this chapter with a discussion on a well-known sequence named after Leonardo of Pisa, who was also called Fibonacci.

Definition. The *Fibonacci sequence* is given recursively by $f_0 = 0$, $f_1 = 1$, and $f_n = f_{n-1} + f_{n-2}$ for $n \geq 2$. The numbers appearing in this sequence, i.e.,

$$0, 1, 1, 2, 3, 5, 8, 13, 21, 34, 55, 89, 144, \ldots$$

are called the *Fibonacci numbers*. Hence, in this notation, f_n denotes the nth Fibonacci number.

The characteristic equation for f_n is $x^2 - x - 1 = 0$, which has distinct roots $x = \frac{1}{2}(1 \pm \sqrt{5})$. With the given f_0 and f_1, Theorem 4.16 leads us to the explicit formula for f_n—strangely involving the irrational number $\sqrt{5}$.

Theorem 4.21. The nth Fibonacci number, $n \geq 0$, is given by

$$f_n = \frac{1}{\sqrt{5}} \left(\frac{1 + \sqrt{5}}{2} \right)^n - \frac{1}{\sqrt{5}} \left(\frac{1 - \sqrt{5}}{2} \right)^n$$

Moreover, the generating function for f_n is given by

$$\sum_{n=0}^{\infty} f_n x^n = \frac{x}{1 - x - x^2}$$

Proof. If $S(x) = \sum f_n x^n$, then $S(x) - xS(x) - x^2 S(x) = f_0 + (f_1 - f_0)x + (f_2 - f_1 - f_0)x^2 + (f_3 - f_2 - f_1)x^3 + \cdots = 0 + x + 0x^2 + 0x^3 + \cdots = x$. This yields the desired generating function. The explicit formula for f_n can then be solved using partial fractions, or directly by Theorem 4.16.   ▽

Exercise* 4.71. For those who know differential equations: Find the exponential generating function $f(x)$ for f_n by considering the first and second derivatives, $f'(x)$ and $f''(x)$.

Exercise 4.72. Show that f_n is the integer nearest to $\frac{1}{\sqrt{5}} \left(\frac{1+\sqrt{5}}{2} \right)^n$ by showing that $\left| \frac{1}{\sqrt{5}} \left(\frac{1-\sqrt{5}}{2} \right)^n \right| < \frac{1}{2}$ for all $n \geq 0$. Hence, we may write

$$f_n = \left\lfloor \frac{1}{\sqrt{5}} \left(\frac{1 + \sqrt{5}}{2} \right)^n + \frac{1}{2} \right\rfloor$$

The Fibonacci sequence satisfies quite a number of other recurrence relations, many of which are elegant to look at and yet readily proved by mathematical induction.

Example. Prove that the Fibonacci numbers f_n satisfy the relation

$$f_0 + f_1 + f_2 + \cdots + f_n = f_{n+2} - 1$$

Solution. We note that $f_0 = f_2 - 1$. Assuming the claim holds, we see that $f_0 + f_1 + f_2 + \cdots + f_{n+1} = (f_{n+2} - 1) + f_{n+1} = f_{n+3} - 1$.

Exercise* 4.73. The Fermat numbers $F_n = 2^{2^n} + 1$, for the sake of analogy, obey the following recurrence relation. Prove this fact for all $n \geq 0$.

$$F_0 \times F_1 \times F_2 \times \cdots \times F_n = F_{n+1} - 2$$

Exercise 4.74. Prove the following identities involving the Fibonacci numbers, where each summation ranges over $0 \leq k \leq n$.

a) $\sum f_{2k+1} = f_{2n+2}$
b) $\sum f_{2k} = f_{2n+1} - 1$
c) $\sum (-1)^k f_k = (-1)^n f_{n-1} - 1$
d) $\sum f_k^2 = f_n f_{n+1}$

Exercise 4.75. Use induction on n to prove that, if $m, n \geq 0$,

$$f_{m+n} = f_m f_{n+1} + f_{m-1} f_n$$

Fibonacci numbers also enjoy some remarkable divisibility properties. We will demonstrate a few of these properties which will lead to the crucial fact that $\gcd(f_m, f_n) = f_{\gcd(m,n)}$.

Example. The nth Fibonacci number f_n is even if and only if n is a multiple of 3. Prove this fact for all $n \geq 0$.

Solution. The Fibonacci sequence begins with $0, 1, 1$, i.e., even, odd, odd. Since $f_n = f_{n-1} + f_{n-2}$, the next three terms will be even (odd plus odd), odd (odd plus even), odd (even plus odd). We see that this pattern continues, giving an even term every multiple of three.

Exercise 4.76. Show that 3 divides f_n if and only if 4 divides n.

Theorem 4.22. If m divides n, then f_m divides f_n.

Proof. We prove the claim by induction, with $n = km$ for $k \geq 2$. Using Exercise 4.75, $f_{2m} = f_{m+m} = f_m f_{m+1} + f_{m-1} f_m$, a multiple of f_m. Assume now that f_m divides f_{km}. Then $f_{(k+1)m} = f_{km+m} = f_{km} f_{m+1} + f_{km-1} f_m$, which is again divisible by f_m as each term is. ▽

Theorem 4.23. If $d = \gcd(m, n)$, then $\gcd(f_m, f_n) = f_d$.

Proof. By Theorem 1.3, it suffices to establish the identity

$$\gcd(f_m, f_n) = \gcd(f_n, f_{m \bmod n})$$

by showing that the two pairs have identical set of common divisors. Recall that $m = qn + m \bmod n$, with the integer $q = \lfloor m/n \rfloor$. By Exercise 4.75,

$$f_m = f_{qn+m \bmod n} = f_{qn} f_{(m \bmod n)+1} + f_{qn-1} f_{m \bmod n}$$

Since f_n divides f_{qn}, any divisor common to f_n and $f_{m \bmod n}$ must divide f_m too. Conversely, let c divide both f_m and f_n. Thus c divides $f_{qn-1} f_{m \bmod n}$. We will be done showing that c divides $f_{m \bmod n}$, which follows if we have $\gcd(f_n, f_{qn-1}) = 1$. The last claim holds for if any number b divides f_{qn-1} and f_n, hence f_{qn}, then b divides $f_{qn-2} = f_{qn} - f_{qn-1}$. By iteration, b will divide $f_{qn-3}, f_{qn-4}, \ldots$, and $f_2 = 1$. ▽

Exercise 4.77. Establish more divisibility properties concerning Fibonacci numbers, stated below.
a) If $f_n > 3$ is prime, then so is n,
b) $\gcd(f_n, f_{n+1}) = 1$ for all $n \geq 0$,
c) f_n divides f_m if and only if n divides m,
d) f_n is a multiple of 5 if and only if n is too.

Exercise* 4.78. Prove that a Fibonacci number f_n has a zero unit digit if and only if n is a multiple of 15.

Fibonacci numbers can also be found in Pascal's triangle. We give two relations involving f_n and $\binom{n}{k}$, the second of which is less obvious but will lead to an interesting compositeness test to complement the ones based on Fermat's little theorem given in Section 1.4.5.

The terms starting at the $\binom{n-1}{0}$ entry in Pascal's triangle (as a lower triangular matrix) going diagonally up, sum to the nth Fibonacci number. For instance, we have $f_7 = \binom{6}{0} + \binom{5}{1} + \binom{4}{2} + \binom{3}{3} = 1 + 5 + 6 + 1 = 13$. In stating the theorem, we remark that $\binom{n}{k} = 0$ is assumed when $k > n$.

Theorem 4.24. For $n \geq 1$, the nth Fibonacci number is given by

$$f_n = \sum_{k \geq 0} \binom{n-k-1}{k}$$

Proof. Let S_n denote the summation on the right. Since $S_1 = S_2 = 1$, it suffices to show that $S_n = S_{n-1} + S_{n-2}$ for $n \geq 3$. Note that

$$S_{n-1} = \binom{n-2}{0} + \binom{n-3}{1} + \binom{n-4}{2} + \binom{n-5}{3} + \cdots$$
$$S_{n-2} = \binom{n-3}{0} + \binom{n-4}{1} + \binom{n-5}{2} + \cdots$$

We use Pascal's formula in adding the two, columnwise, to get

$$S_{n-1} + S_{n-2} = \binom{n-1}{0} + \binom{n-2}{1} + \binom{n-3}{2} + \binom{n-4}{4} + \cdots = S_n$$

where we have replaced $\binom{n-2}{0} = 1$ by $\binom{n-1}{0}$. $\triangledown$

Theorem 4.25. For $n \geq 1$, we have

$$2^{n-1} f_n = \sum_{k \geq 0} \binom{n}{2k+1} 5^k$$

Proof. We use the explicit formula for f_n and the binomial theorem:

$$2^n \sqrt{5} f_n = \left(1 + \sqrt{5}\right)^n - \left(1 - \sqrt{5}\right)^n = 2 \sum \binom{n}{2k+1} \sqrt{5}^{2k+1}$$

Dividing both sides by $2\sqrt{5}$ yields the result. $\triangledown$

Theorem 4.26. Let $p > 5$ be a prime number. Then, either p divides both f_{p-1} and $f_p - 1$, or else p divides both f_{p+1} and $f_p + 1$.

Proof. Let $n = p$ in Theorem 4.25. We have $2^{p-1} \bmod p = 1$ by Fermat's little theorem, whereas Theorem 4.14 says that $\binom{p}{k} \bmod p = 0$, unless $k = 0$ or $k = p$. Hence $f_p \equiv 5^{(p-1)/2} \pmod{p}$. Since now $f_p^2 \equiv 5^{p-1} \equiv 1 \pmod{p}$, by Theorem 1.14, we must have $f_p \equiv \pm 1 \pmod{p}$.

A similar result holds with $n = p + 1$ in Theorem 4.25, where $2f_{p+1} \equiv 1 + 5^{(p-1)/2} \equiv 1 + f_p \pmod{p}$. In particular, p divides f_{p+1} if and only if $f_p \equiv -1 \pmod{p}$.

As for $n = p - 1$, see Exercise 4.49, which claims that each binomial coefficient appearing in Theorem 4.25 is congruent to $-1 \pmod{p}$. In that case,

$$-2^{p-2} f_{p-1} \equiv 1 + 5 + 5^2 + \cdots + 5^{(p-3)/2} \pmod{p}$$

And multiplying both sides by -4 gives us $2f_{p-1} \equiv 1 - 5^{(p-1)/2} \pmod{p}$. This time, p divides f_{p-1} if and only if $f_p \equiv 1 \pmod{p}$. Since we have stated that $f_p \equiv \pm 1 \pmod{p}$, the proof is complete. $\triangledown$

The contrapositive of Theorem 4.26 can be used to recognize a large composite n which fails the given divisibility properties. To compute f_n efficiently, we need a matrical version of the successive squaring algorithm (Section 1.4.3) based on the fact that, for $n \geq 1$,

$$\begin{bmatrix} f_{n+1} & f_n \\ f_n & f_{n-1} \end{bmatrix} = \begin{bmatrix} 1 & 1 \\ 1 & 0 \end{bmatrix}^n \tag{4.6}$$

Exercise 4.79. Prove the matrical identity (4.6) using induction.

Example. Use the Fibonacci sequence to show that 323 is a composite.

Solution. We wish to compute the identity (4.6) for $n = 323$. This is done by successively squaring the matrix on the right, each time reducing the entries mod 323:

$$\begin{bmatrix} 1 & 1 \\ 1 & 0 \end{bmatrix}^2 = \begin{bmatrix} 2 & 1 \\ 1 & 1 \end{bmatrix} \qquad \begin{bmatrix} 1 & 1 \\ 1 & 0 \end{bmatrix}^4 = \begin{bmatrix} 5 & 3 \\ 3 & 2 \end{bmatrix}$$

$$\begin{bmatrix} 1 & 1 \\ 1 & 0 \end{bmatrix}^8 = \begin{bmatrix} 34 & 21 \\ 21 & 13 \end{bmatrix} \qquad \begin{bmatrix} 1 & 1 \\ 1 & 0 \end{bmatrix}^{16} = \begin{bmatrix} 305 & 18 \\ 18 & 287 \end{bmatrix}$$

$$\begin{bmatrix} 1 & 1 \\ 1 & 0 \end{bmatrix}^{32} = \begin{bmatrix} 2 & 320 \\ 320 & 5 \end{bmatrix} \qquad \begin{bmatrix} 1 & 1 \\ 1 & 0 \end{bmatrix}^{64} = \begin{bmatrix} 13 & 302 \\ 302 & 34 \end{bmatrix}$$

$$\begin{bmatrix} 1 & 1 \\ 1 & 0 \end{bmatrix}^{128} = \begin{bmatrix} 287 & 305 \\ 305 & 305 \end{bmatrix} \qquad \begin{bmatrix} 1 & 1 \\ 1 & 0 \end{bmatrix}^{256} = \begin{bmatrix} 5 & 3 \\ 3 & 2 \end{bmatrix}$$

Since $323 = 101000011_2 = 256 + 64 + 2 + 1$, we then have, mod 323,

$$\begin{bmatrix} f_{324} & f_{323} \\ f_{323} & f_{322} \end{bmatrix} = \begin{bmatrix} 5 & 3 \\ 3 & 2 \end{bmatrix} \begin{bmatrix} 13 & 302 \\ 302 & 34 \end{bmatrix} \begin{bmatrix} 2 & 1 \\ 1 & 1 \end{bmatrix} \begin{bmatrix} 1 & 1 \\ 1 & 0 \end{bmatrix} = \begin{bmatrix} 0 & 1 \\ 1 & 322 \end{bmatrix}$$

We note that 323 divides f_{324} but not $f_{323} + 1$, and 323 divides $f_{323} - 1$, but not f_{322}. Failing Theorem 4.26 test, 323 is definitely composite.

It should be pointed out that the converse of Theorem 4.26 does not hold in general, for there are composites which well pass this test.

Exercise 4.80. Four composites are given below. Determine which ones will go undetected by Theorem 4.26.

a) 377
b) 1891
c) 3827
d) 4181

In particular, a weaker contrapositive of Theorem 4.26 has been used to define a new family of pseudoprimes, in conjunction with the fact that

$$5^{\frac{p-1}{2}} \equiv \begin{cases} +1 & (\text{mod } p) \quad \text{if} \quad p \equiv \pm 1 \quad (\text{mod } 10) \\ -1 & (\text{mod } p) \quad \text{if} \quad p \equiv \pm 3 \quad (\text{mod } 10) \end{cases}$$

It is not our intention here, however, to pursue the necessary theoretical tools in order to establish these claims.

Definition. A *Fibonacci pseudoprime* is a composite n of unit digit 1 or 9, such that n divides f_{n-1}, or a composite n of unit digit 3 or 7, such that n divides f_{n+1}.

The composite $n = 323$ in the previous example is therefore a Fibonacci pseudoprime (the smallest one, in fact) since f_{324} is divisible by 323.

Exercise 4.81. Find more examples of Fibonacci pseudoprimes among the composites given in Exercise 4.80.

As a final remark, we note the connection between Fibonacci numbers and the famous golden ratio.

Definition. The *golden ratio* is an irrational constant φ defined by

$$\varphi = \frac{1 + \sqrt{5}}{2} = 1.6180339887\ldots$$

(The symbol φ is a variation of the lowercase Greek letter *phi*.) Recall that φ is one root of the characteristic equation $x^2 - x - 1 = 0$ for the Fibonacci sequence—the other root being $1 - \varphi = -1/\varphi$.

Question. Can you recall two other variants of *phi* used in this text?

It was the ancient Greeks who gave the golden ratio its name, perhaps for its fascinating appearance in nature as the ratio of two quantitites a and b, for which $a/b = (a+b)/a$. You will check that such a relation can hold if and only if $a/b = \varphi$. To visualize, with the fact that

$$\lim_{n\to\infty} \frac{f_{n+1}}{f_n} = \varphi$$

the sequence of nested rectangles of size $f_n \times f_{n+1}$ depicted below, increasingly approaches the *golden rectangle,* i.e., one whose length-to-width ratio is given by φ.

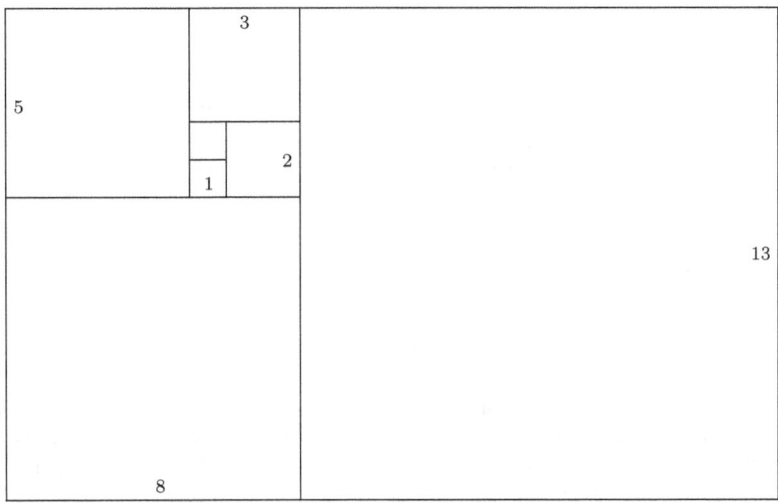

Books to Read

1. T. Andreescu and Z. Feng, *A Path to Combinatorics for Undergraduates: Counting Strategies*, Birkhäuser 2003.

2. R. A. Dunlap, *The Golden Ratio and Fibonacci Numbers*, World Scientific 1998.

3. A. W. F. Edwards, *Pascal's Arithmetical Triangle: The Story of a Mathematical Idea*, Johns Hopkins University Press 2002.

4. H. Gordon, *Discrete Probability*, Springer 1997.

Chapter 5

Topics in Graph Theory

Graphs are a quite common mathematical term which we may encounter in simple geometry or Calculus, and particularly we have seen them as the digraphs of binary relations in Section 3.1.2. Modern graph theory, however, has developed into a very advanced study which has many powerful applications. This chapter is but a shallow survey into some of the most familiar topics in the subject.

5.1   Some Basic Features

Definition. A *graph* G is a composite of a finite set V_G of *vertices* and another set E_G of *edges*, where an edge is defined to be a set of two distinct vertices. When there is no ambiguity, we write V and E, instead of V_G and E_G, respectively. If $E = \emptyset$, we have a *trivial graph* of only vertices.

A graph can be represented by a picture in very much the same way we draw vertices and edges of a digraph. For example, the picture given below

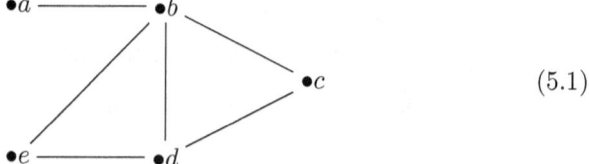

$$(5.1)$$

represents the graph G with the set of vertices $V = \{a, b, c, d, e\}$ and edges $E = \{\{a,b\}, \{b,c\}, \{b,d\}, \{b,e\}, \{c,d\}, \{d,e\}\}$. Unlike digraphs, however, note that here an edge has no direction—thus we use set notation $\{a,b\}$ in place of the ordered pair (a,b) to represent an edge.

5.1.1 Degrees

Definition. For brevity, we denote the edge $\{a, b\} \in E$ simply by ab, if doing so does not cause misunderstanding. In particular when $ab \in E$, we say that the vertices a and b are *adjacent*. We define the *degree* of a vertex a, written $\deg(a)$, to be the number of vertices in G which are adjacent to a. Moreover, let $\deg G = \sum_{a \in V_G} \deg(a)$, the *degree* of the graph G.

In our example of G given by (5.1) earlier, we have $\deg(a) = 1$, $\deg(b) = 4$, $\deg(c) = 2$, $\deg(d) = 3$, $\deg(e) = 2$, and $\deg G = 1 + 4 + 2 + 3 + 2 = 12$.

Theorem 5.1 (Euler's Theorem). The degree of any graph G is twice the number of its edges, i.e., $\deg G = 2\,|E_G|$. In particular, the degree of any graph is an even number.

Proof. Every edge $ab \in E$ contributes two toward the degree sum, one for $\deg(a)$ and another for $\deg(b)$, thus the result. $\triangledown$

If we replace E_G by a multiset, allowing repetition of edges, then G will be called a *multigraph*. Sometimes a vertex in a multigraph is allowed to be adjacent to itself, creating a *loop*, i.e., an element $aa \in E$. Multigraphs, however, are not a concern in this text.

Exercise* 5.1. Show that Euler's theorem holds for multigraphs as well, if $\deg(a)$ is defined to be the number of edges containing a, counting multiplicity and duplicity of each loop.

Definition. The following four families of particularly interesting and useful graphs will each be given a special name.

1) A *complete* graph K_n is a graph with n vertices, all of which are adjacent one to another. In particular, K_3 is also called a *triangle*.

2) The $m + n$ vertices of the graph $K_{m,n}$ are bipartitioned—i.e., partitioned into two subsets—into m and n elements each, such that two vertices are adjacent if and only if they do not belong together. A graph with this property is called *complete bipartite*.

3) A *path* P_n is a graph with n vertices, $V = \{v_1, v_2, \ldots, v_n\}$, and $n - 1$ edges, $E = \{v_1 v_2, v_2 v_3, \ldots, v_{n-1} v_n\}$. With this notation, we say that P_n is a path from v_1 to v_n.

4) For $n \geq 3$, let $C_n = P_n \cup \{v_n v_1\}$, which we call a *closed path* or a *cycle* of n vertices.

Question. Can you name five mathematical terms given throughout this text, one in each chapter, which begin with the prefix *bi-*?

We illustrate below possible drawings of some of these special graphs.

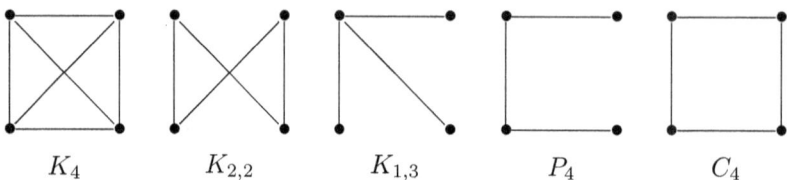

K_4 $K_{2,2}$ $K_{1,3}$ P_4 C_4

Question. Can you spot two graphs among the five pictured above, which are essentially identical sets of vertices and edges?

Exercise 5.2. Evaluate the degree of each graph K_n, $K_{m,n}$, P_n, and C_n, and count the number of edges in each graph as well.

Exercise* 5.3. Prove that the number of edges in $K_2, K_3, K_4, \ldots$ form the sequence of the *triangular numbers,* i.e., $1, 1+2, 1+2+3, \ldots$

Test 5.4. Which one of these four graphs has the most number of edges?

a) K_{99}
b) $K_{50,50}$
c) P_{100}
d) C_{200}

Test 5.5. The degree of a complete bipartite graph is 210. What could be the smallest number of vertices this graph has?

a) 15
b) 22
c) 26
d) 38

Definition. The *Peterson graph* consists of two copies of C_5 which are interconnected to each other in the way depicted below.

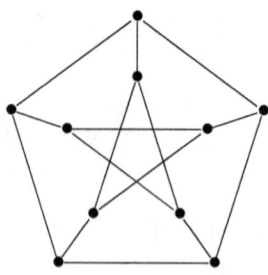

Question. What is the degree of the Peterson graph?

Definition. A graph is called *regular* if all its vertices have equal degrees, otherwise *irregular*. In particular, a regular graph in which $\deg(a) = d$ for every vertex a, is called *d-regular*. For example, K_3 is 2-regular, whereas P_5 is irregular.

Question. Is the Peterson graph regular?

Test 5.6. Which one of these four graphs is not regular?
a) K_{99}
b) $K_{99,99}$
c) P_{99}
d) C_{99}

Exercise 5.7. Analyze the regularity of the graphs K_n, $K_{m,n}$, P_n, and C_n.

Definition. If $V_G = \{v_1, v_2, \ldots, v_n\}$, the *degree sequence* of the graph G is the sequence $\{\deg(v_i)\}$ of length n, arranged in decreasing order. For example, the degree sequence of P_5 is $2, 2, 2, 1, 1$.

Question. What is the degree sequence of the Peterson graph?

Exercise 5.8. Find the degree sequence of each K_n, $K_{m,n}$, P_n, and C_n.

We call a degree sequence such as $2, 2, 2, 1, 1$ *graphical* since there exists a graph, e.g., P_5, having this degree sequence. Obviously, not all degree sequences are graphical; for instance, $4, 3, 2, 2, 2$ cannot possibly be graphical, says Euler's theorem. (Why?) The following algorithm can be used to determine whether a given degree sequence is graphical.

Question. Why is $4, 3, 2, 1$ not graphical either?

Algorithm 5.2 (Graphical Degree Sequence). Given a degree sequence, we determine graphical or not graphical.

1) Delete the first integer, say k.

2) From what remains, subtract the first k numbers each by 1. If this is not possible, the sequence is not graphical. If we get all zeros, the sequence is graphical.

3) If necessary, rearrange the newly formed sequence in decreasing order, and repeat the above steps until a conclusion is obtained.

Observation. In each round, if the resulting new sequence is graphical, so is the preceding one, by simply restoring the deleted vertex and its associated edges. To prove the converse is not that trivial, though. ▽

Example. Determine if $3, 2, 2, 1, 1, 1$ is a graphical degree sequence.

Solution. Remove the first number, 3, and subtract the first three numbers of what remains each by 1. We get $1, 1, 0, 1, 1$. The complete iterations recorded below show that the sequence is graphical.

$$3, 2, 2, 1, 1, 1 \rightarrow 1, 1, 0, 1, 1 \rightarrow 1, 1, 1, 1, 0 \rightarrow 0, 1, 1, 0 \rightarrow 1, 1, 0, 0 \rightarrow 0, 0, 0$$

Exercise 5.9. Apply the algorithm to these sequences. If a sequence is graphical, draw a graph satisfying the given degree sequence.
a) $3, 2, 2, 1, 1, 1$
b) $4, 3, 3, 2, 1, 0$
c) $5, 3, 2, 2, 1, 1$
d) $5, 4, 4, 3, 3, 3, 3, 2, 2, 1$

Exercise* 5.10. For multigraphs, show that a degree sequence is graphical if and only they sum to an even number.

5.1.2   Isomorphisms

Definition. Two graphs are *isomorphic* to each other, written $G \simeq H$, if there is a bijection $f : V_G \rightarrow V_H$ such that a and b are adjacent in G if and only if $f(a)$ and $f(b)$ are adjacent in H.

For example, $K_3 \simeq C_3$, so both are called triangles. Note also that $K_2 \simeq K_{1,1} \simeq P_2$. Being isomorphic means that the two graphs can be represented by identical pictures, if we discount the labeling of the vertices.

Test 5.11. Which one of these graph isomorphisms is false?
a) $K_{1,2} \simeq P_3$
b) $K_{1,3} \simeq P_4$
c) $K_{2,2} \simeq C_4$
d) All the above are true statements.

If $G \simeq H$, then clearly G and H must have the same number of vertices, the same number of edges, and identical degree sequences. However, it is important to note that none of these properties is a sufficient condition for isomorphism.

Example. These two graphs given below have each 6 vertices, 5 edges, and degree sequence $3, 2, 2, 1, 1, 1$. Prove that they are not isomorphic to each other.

Solution. Any bijection between them must preserve the unique vertex of degree 3. Observe that on the left, that unique vertex is adjacent to three others with degrees $2, 1, 1$. On the right, however, the adjacent three have degrees $2, 2, 1$. This proves that no such bijection will work.

Question. Is there another graph, isomorphic to neither of the above two, but with the same degree sequence $3, 2, 2, 1, 1, 1$?

Exercise 5.12. Find several (two to four) non-isomorphic graphs with each given degree sequence.

a) $4, 4, 3, 2, 2, 1$
b) $2, 2, 2, 2, 2, 1, 1$
c) $5, 3, 2, 2, 1, 1, 1, 1$
d) $3, 3, 3, 3, 3, 3, 3, 3$

Definition. A graph H is a *subgraph* of the graph G if $V_H \subseteq V_G$ and $E_H \subseteq E_G$. We denote this relation by $H \subseteq G$, and say that G *contains* H.

By abuse of notation, we may write $H \subseteq G$ when we really mean that G contains a subgraph which is isomorphic to H. Hence, for example, $P_3 \subseteq C_3$ and $K_{2,2} \subseteq K_{2,4}$.

Test 5.13. Which one of these four graphs does not contain C_4?

a) K_5
b) $K_{5,5}$
c) $K_{3,5}$
d) C_5

Question. How many non-isomorphic cycles are contained in the Peterson graph?

Definition. A graph G is called *connected* if there is a path from any vertex to any other vertex in G, otherwise *disconnected*. A *component* of G is then a maximal connected subgraph of G.

Hence, a graph is disconnected if and only it has more than one component. It is clear that isomorphic graphs must have the same number of components.

Exercise 5.14. If G is connected, prove that $|E_G| \geq |V_G| - 1$.

Definition. An edge in a connected graph G is a *bridge* when G would become disconnected if this edge be removed.

Question. Does the Peterson graph have a bridge?

Test 5.15. Which one of these four graphs contains a bridge?

a) K_9
b) $K_{2,9}$
c) P_9
d) C_9

Exercise 5.16. Determine exactly when each of the graphs K_n, $K_{m,n}$, P_n, and C_n, may contain a bridge.

Definition. The *complement* of a graph G is the graph $\overline{G}$, where $V_{\overline{G}} = V_G$ and $ab \in E_{\overline{G}}$ if and only if $ab \notin E_G$.

Below we show the picture of C_5 next to its complement.

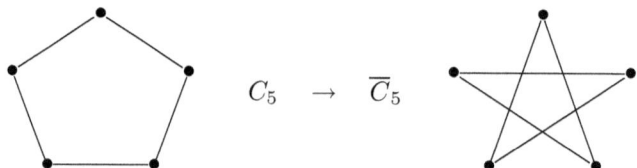

$$C_5 \quad \rightarrow \quad \overline{C}_5$$

Exercise 5.17. Draw the complements of P_4, C_4, K_5, and of $K_{5,4}$.

Question. What is the degree of the complement of the Peterson graph?

Theorem 5.3. If G is disconnected, then $\overline{G}$ is connected.

Proof. Let $a, b \in V_{\overline{G}}$. If a and b are not adjacent in G, then they are in $\overline{G}$. If $ab \in E_G$, choose a vertex c in G belonging to a component different than that containing a and b. Such a vertex c exists if G is disconnected. Then $ac \notin E_G$ and $cb \notin E_G$, and they form a path from a to b in $\overline{G}$.   ▽

Definition. A graph G is *self-complementary* when $\overline{G} \simeq G$.

In fact, you have seen a self-complementary graph (where?) i.e., $C_5 \simeq \overline{C}_5$. Another example, P_4 is also self-complementary—a fact which is not hard to sketch and see. Note that Theorem 5.3 requires a self-complementary graph to be connected.

Exercise 5.18. Find one more example of a self-complementary graph. If it helps, see the next exercise first.

Exercise* 5.19. Prove that a self-complementary graph with n vertices must have exactly $\frac{n(n-1)}{4}$ edges and consequently, $n \bmod 4 \leq 1$.

5.1.3   Matrices

If M denotes an arbitrary matrix, we shall use the notation $[M]_{ij}$ to refer
to the (i, j) entry in M, i.e., the entry in the ith row and jth column of M.

Definition. Suppose that $V_G = \{v_1, v_2, \ldots, v_n\}$. The *adjacency matrix* of
the graph G is the $n \times n$ matrix A given by $[A]_{ij} = 1$ if $v_i v_j \in E_G$, otherwise
$[A]_{ij} = 0$.

For example, the adjacency matrix of C_4—with the standard edges $E =$
$\{v_1 v_2, v_2 v_3, v_3 v_4, v_4 v_1\}$—is given by

$$
A = \begin{bmatrix} 0 & 1 & 0 & 1 \\ 1 & 0 & 1 & 0 \\ 0 & 1 & 0 & 1 \\ 1 & 0 & 1 & 0 \end{bmatrix}
$$

Question. Can non-isomorphic graphs have the same adjacency matrix?

Exercise 5.20. Describe the adjacency matrices of the graphs K_n, $K_{m,n}$,
P_n, and C_n.

Exercise* 5.21. Let A denote the adjacency matrix of a graph G with
vertices $v_1, v_2, \ldots, v_n$. Prove that $[A^2]_{ii} = \deg(v_i)$.

Definition. A *permutation matrix* is a square matrix obtained from the
identity matrix by reordering its rows.

For example, the matrix

$$
P = \begin{bmatrix} 1 & 0 & 0 & 0 & 0 \\ 0 & 0 & 1 & 0 & 0 \\ 0 & 0 & 0 & 0 & 1 \\ 0 & 1 & 0 & 0 & 0 \\ 0 & 0 & 0 & 1 & 0 \end{bmatrix} \tag{5.2}
$$

is a permutation matrix, which we may obtain from the identity matrix by
permuting its rows in the order of $(1, 3, 5, 2, 4)$. A known fact is that every
permutation matrix P belongs to the family of orthogonal matrices, as we
have $P^{-1} = P^T$.

Since isomorphic graphs are essentially relabeling of the vertices, we can
expect that their adjacency matrices are related to each other via some permutation matrix. The following theorem, though not quite useful, describes
this relation in an elegant manner.

Theorem 5.4. Suppose that A and B are the adjacency matrices of the graphs G and H, respectively. Then $G \simeq H$ if and only if $B = PAP^T$ for some permutation matrix P.

Example. We look again at the fact that $C_5 \simeq \overline{C}_5$, this time with the particular vertex labeling as follows.

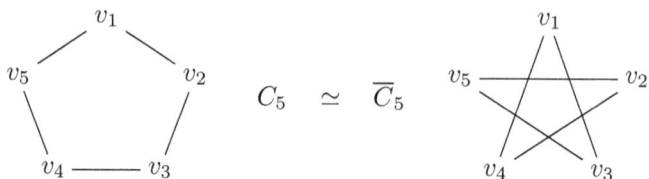

An obvious bijection is one which permutes the vertex indices $(1, 2, 3, 4, 5)$ to $(1, 3, 5, 2, 4)$, thus the permutation matrix P given in (5.2). Indeed, the corresponding adjacency matrices A and $\overline{A}$, respectively, satisfy the relation $PAP^T = \overline{A}$, i.e.,

$$\begin{bmatrix} 1 & 0 & 0 & 0 & 0 \\ 0 & 0 & 1 & 0 & 0 \\ 0 & 0 & 0 & 0 & 1 \\ 0 & 1 & 0 & 0 & 0 \\ 0 & 0 & 0 & 1 & 0 \end{bmatrix} \begin{bmatrix} 0 & 1 & 0 & 0 & 1 \\ 1 & 0 & 1 & 0 & 0 \\ 0 & 1 & 0 & 1 & 0 \\ 0 & 0 & 1 & 0 & 1 \\ 1 & 0 & 0 & 1 & 0 \end{bmatrix} \begin{bmatrix} 1 & 0 & 0 & 0 & 0 \\ 0 & 0 & 0 & 1 & 0 \\ 0 & 1 & 0 & 0 & 0 \\ 0 & 0 & 0 & 0 & 1 \\ 0 & 0 & 1 & 0 & 0 \end{bmatrix} = \begin{bmatrix} 0 & 0 & 1 & 1 & 0 \\ 0 & 0 & 0 & 1 & 1 \\ 1 & 0 & 0 & 0 & 1 \\ 1 & 1 & 0 & 0 & 0 \\ 0 & 1 & 1 & 0 & 0 \end{bmatrix}$$

Definition. Suppose that $V_G = \{v_1, v_2, \ldots, v_n\}$ and $E_G = \{e_1, e_2, \ldots, e_m\}$. Then the *incidence matrix* of the graph G is the $n \times m$ matrix Z given by $[Z]_{ij} = 1$ if $v_i \in e_j$, otherwise $[Z]_{ij} = 0$.

For example, the incidence matrix of the path P_4, with standard set of edges $E = \{v_1v_2, v_2v_3, v_3v_4\}$, is the following 4×3 matrix Z.

$$Z = \begin{bmatrix} 1 & 0 & 0 \\ 1 & 1 & 0 \\ 0 & 1 & 1 \\ 0 & 0 & 1 \end{bmatrix}$$

Exercise 5.22. Describe the incidence matrices of the graphs K_n, $K_{m,n}$, P_n, and Z_n.

Exercise 5.23. Convert the following incidence matrices to adjacency matrices.

a) $\begin{bmatrix} 1 & 0 \\ 0 & 1 \\ 1 & 1 \end{bmatrix}$
b) $\begin{bmatrix} 0 & 1 & 1 \\ 1 & 0 & 1 \\ 1 & 1 & 0 \end{bmatrix}$
c) $\begin{bmatrix} 1 & 0 & 0 & 0 \\ 1 & 1 & 1 & 0 \\ 0 & 1 & 0 & 1 \\ 0 & 0 & 1 & 1 \end{bmatrix}$
d) $\begin{bmatrix} 1 & 0 & 0 & 0 & 0 & 1 & 1 & 0 \\ 1 & 0 & 1 & 0 & 1 & 0 & 0 & 0 \\ 0 & 1 & 0 & 1 & 1 & 0 & 0 & 0 \\ 0 & 1 & 1 & 0 & 0 & 0 & 1 & 1 \\ 0 & 0 & 0 & 1 & 0 & 1 & 0 & 1 \end{bmatrix}$

Exercise 5.24. Amira has correctly computed the adjacency matrix of an unlabeled graph to be

$$A = \begin{bmatrix} 0 & 1 & 1 & 1 \\ 1 & 0 & 0 & 1 \\ 1 & 0 & 0 & 0 \\ 1 & 1 & 0 & 0 \end{bmatrix}$$

Elias, using a labeling of his own, is to find the incidence matrix. Which one of the following matrices is supposedly Elias's correct answer?

a) $\begin{bmatrix} 1 & 1 & 0 & 0 \\ 1 & 0 & 1 & 0 \\ 0 & 0 & 1 & 1 \\ 0 & 1 & 0 & 1 \end{bmatrix}$ b) $\begin{bmatrix} 1 & 1 & 0 & 0 \\ 1 & 0 & 1 & 0 \\ 0 & 1 & 1 & 1 \\ 0 & 0 & 0 & 1 \end{bmatrix}$ c) $\begin{bmatrix} 0 & 0 & 1 & 1 \\ 0 & 0 & 1 & 1 \\ 1 & 1 & 0 & 0 \\ 1 & 1 & 0 & 0 \end{bmatrix}$ d) $\begin{bmatrix} 1 & 1 & 1 & 1 \\ 1 & 0 & 0 & 0 \\ 0 & 1 & 0 & 1 \\ 0 & 0 & 1 & 0 \end{bmatrix}$

Definition. Let G be a graph with vertices $v_1, v_2, \ldots, v_n$. The *degree matrix* of G is the $n \times n$ diagonal matrix D given by $[D]_{ii} = \deg(v_i)$.

Question. Can you describe the degree matrix of the Peterson graph?

Theorem 5.5. Suppose that the adjacency matrix A and incidence matrix Z have been given for the same graph G. Then $ZZ^T = A + D$, where D is the degree matrix of G.

Proof. For $i \neq j$, we have $[ZZ^T]_{ij} = \sum_{k \geq 1} [Z]_{ik}[Z]_{jk}$. If $v_i v_j \in E$, then there is exactly one value of k for which $[Z]_{ik} = [Z]_{jk} = 1$. In that case, both ZZ^T and $A + D$ have 1 in their (i, j) entries. If $v_i v_j \notin E$, no such k exists, and that entry will be 0 in both. Lastly, if $i = j$, then $[ZZ^T]_{ii} = \sum_{k \geq 1} [Z]_{ik}$, counting the number of vertices adjacent to v_i, which agrees with the ith diagonal entry in $A + D$. $\triangledown$

5.2   Introduction to Trees

Definition. A *tree* is a connected graph which contains no cycles. In general, a graph which contains no cycles is called *acyclic*.

For example, K_4 contains the cycle C_4 (as well as C_3), hence K_4 is not a tree. On the other hand, P_4 is a tree because no subgraph of P_4 is a cycle.

Test 5.25. Which one of these four graphs is a tree?
a) K_9
b) $K_{9,1}$
c) $K_{9,9}$
d) C_9

Exercise 5.26. Determine exactly when each of the graphs K_n, $K_{m,n}$, P_n, and C_n, is a tree.

Question. Is the Peterson graph a tree?

Exercise 5.27. Draw all non-isomorphic trees with no more than 6 vertices.

Theorem 5.6. Let G be a connected graph. The following statements are equivalent one to another.

1) The graph G is acyclic.

2) Every edge in G is a bridge.

3) The size of G is determined by $|E_G| = |V_G| - 1$.

4) There is a unique path between any two vertices in G.

Proof. Suppose that from a to b we have two distinct paths. The union of the two paths make a cycle. A cycle contains no bridge, for removing an edge results in a path through all existing vertices. Now removing a non-bridge edge, if any, keeps the graph connected, hence $|E| > |V| - 1$ by Exercise 5.14. We have established $\neg(4) \rightarrow \neg(1) \rightarrow \neg(2) \rightarrow \neg(3)$, and we leave it to you now to complete the proof of $\neg(3) \rightarrow \neg(4)$. $\triangledown$

Exercise 5.28. Prove that adding an edge to a tree will produce a cycle.

Definition. In a graph, a vertex of degree one is called a *leaf.*

Exercise 5.29. Show that any one of the special graphs K_n, $K_{m,n}$, P_n, and C_n, may have a leaf if and only if it is a tree.

Theorem 5.7. Every tree has a leaf. In fact, a tree has $2 + \sum_{i \geq 3}(i - 2)n_i$ leaves, where n_i denotes the number of vertices of degree i.

Proof. Let $d_1, d_2, \ldots, d_n$ be the degree sequence of a tree. Being connected, we have $d_n \geq 1$. With $n - 1$ edges, Euler's theorem gives $\sum d_i = 2n - 2$. The number of leaves is minimum in the case $2, 2, \ldots, 2, 1, 1$. A vertex of degree $i \geq 3$ reduces this number of twos, while increases the number of leaves—exactly $i - 2$ of them. This yields the claimed formula. $\triangledown$

Exercise* 5.30. So the degree sequence $d_1, d_2, \ldots, d_n$ of a tree, with $d_n > 0$, always sum to $2n - 2$. Conversely, show that every degree sequence with this property is graphical and can be given by a tree.

Definition. A *spanning tree* of a graph G is a tree $T \subseteq G$ with $V_T = V_G$.

For example, P_4 is a spanning tree of both C_4 and K_4. Alternately, a spanning tree for K_4 can be $K_{1,3}$.

Question. Can you find a spanning tree for the Peterson graph in the form of a path?

Test 5.31. Which one of the following graphs is a spanning tree of the graph given in (5.3) below?

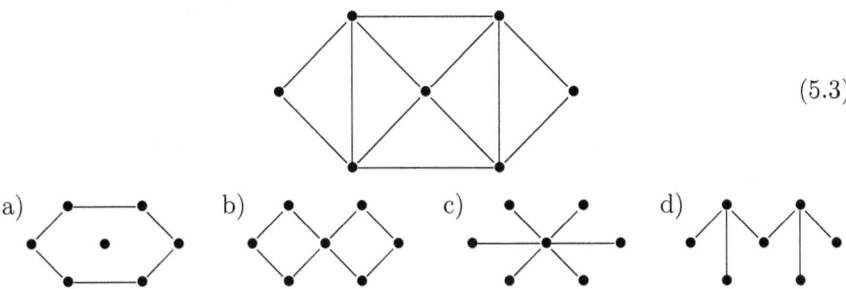

$$(5.3)$$

The following algorithm can be used to generate a spanning tree of a given graph, if connected. If G is disconnected, of course, G cannot have a spanning tree.

Algorithm 5.8 (Depth-First Search). Given a graph G, we determine whether or not G is connected, in which case, we produce a spanning tree of G. We shall label the vertices of G as $v_1, v_2, v_3, \ldots$, based upon the order of traversal as follows.

1) Start with an arbitrary vertex and label this vertex v_1.

2) Move to any vertex adjacent to the current selection which has not been labeled. If no such vertex exists, backtrack to a previously visited vertex.

3) Repeat the previous step until no more move is possible.

4) If all vertices of G have been labeled, then G is connected and this traversal generates a spanning tree of G.

Observation. The algorithm works simply because G is connected if and only if every other vertex will eventually be traversed, regardless the choice of v_1. That the resulting graph is a tree is due to the fact that it takes exactly one new edge to add one new vertex during the entire procedure. ▽

In general, a graph can have many spanning trees, unless G is already a tree, in which case G has no spanning tree other than G itself. The next theorem provides an algorithm to compute the number of spanning trees of a graph with labeled vertices. The proof, first given by Kirchoff, can be found in many graph theory texts.

Theorem 5.9 (Matrix Tree Theorem). Let G be a connected graph with labeled vertices, adjacency matrix A, and degree matrix D. Then any cofactor of the matrix $D - A$ will give the number of spanning trees of G.

Example. The graph G is given, together with the associated matrix $D - A$.

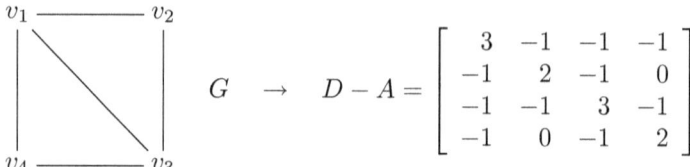

$$G \quad \rightarrow \quad D - A = \begin{bmatrix} 3 & -1 & -1 & -1 \\ -1 & 2 & -1 & 0 \\ -1 & -1 & 3 & -1 \\ -1 & 0 & -1 & 2 \end{bmatrix}$$

Recall that the cofactor $C_{i,j}$ of a square matrix M is the determinant of the matrix obtained from M by removing the ith row and jth column, e.g.,

$$C_{3,1} = \det \begin{bmatrix} -1 & -1 & -1 \\ 2 & -1 & 0 \\ 0 & -1 & 2 \end{bmatrix}$$

We compute this determinant to conclude that there are 8 spanning trees for the graph G.

Exercise 5.32. Repeat the example using P_4, K_4, $K_{2,3}$, and C_5.

Exercise* 5.33. There are n^{n-2} different trees with vertices $v_1, v_2, \ldots, v_n$, if $n \geq 2$. Prove this claim using the matrix tree theorem by showing that n^{n-2} is the number of spanning trees of a labeled K_n.

Suppose the vertices represent towns in a state whose government wishes to build railroads connecting all of them. To save cost, it would be wise to choose not only a spanning tree, but one whose edges sum to the least possible kilometers. This is one application of a minimal spanning tree.

Definition. A graph is *weighted* if every edge is associated with a numerical value, called *weight*. For us, weighted graphs are allowed only non-negative values. A *minimal spanning tree* of a weighted graph is a spanning tree with the least total weight.

Question. What is the least possible total weight for a spanning tree of the weighted graph (5.4) below?

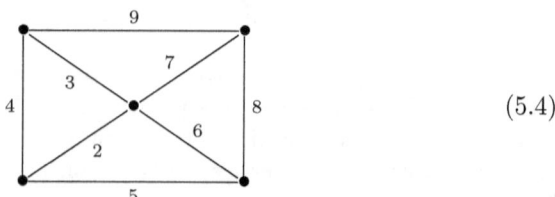

$$(5.4)$$

In order to present the algorithm by which we can find a minimal spanning tree of a weighted graph, we first need a definition.

Definition. For every vertex $v \in V$, let $N(v) = \{w \in V \mid vw \in E\}$, i.e., the set of all vertices which are adjacent to v. Now if $S \subseteq V$, we define the *neighborhood* of S to be the set $N(S) = \bigcup_{v \in S} N(v)$.

Algorithm 5.10 (Prim). Given a connected weighted graph, we produce a minimal spanning tree as follows.

1) Select an arbitrary vertex. Let S be the set of vertices which have already been selected.

2) Add to S one more vertex from $N(S)$ such that the corresponding edge is of least weight.

3) Repeat until $|S| = |V|$.

Observation. It is clear that the resulting subgraph is a tree since we would have selected exactly $|V|$ vertices and $|V| - 1$ edges, keeping the subgraph connected at each selection. Although minimal spanning tree may not be unique, it can be shown that this algorithm always yields the least possible total weight. $\triangledown$

Exercise 5.34. Apply Prim's algorithm to the weighted graph below.

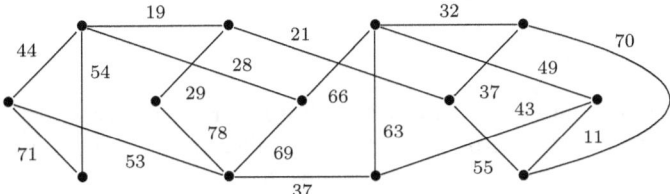

5.3 Walks

Definition. A *walk* is simply a sequence of continuous edges, i.e., of the form $v_1 v_2, v_2 v_3, \dots, v_n v_{n+1}$. In this case, it is a walk of *length* n from v_1 to v_{n+1}. If it happens that $v_{n+1} = v_1$, we have what we call a *closed walk* of length n. We loosely treat a walk as a graph containing these edges and the vertices therein.

In particular, a path P_n can be considered as a walk of length $n - 1$, while C_n a closed walk of length n. In a walk, the edges involved are not assumed distinct, and neither are the vertices—unless of course, the walk is a path.

Theorem 5.11. If G is a closed walk of odd length, then G contains a cycle of odd length.

Proof. A closed walk of length three is none other than C_3, so the claim is true. We proceed by induction, assuming the theorem has been proved for all lenghts less than n. If G is the walk $v_1v_2, v_2v_3, \ldots, v_nv_1$ with no repeated vertex, then $G \simeq C_n$ and we are done. Suppose now $v_i = v_{i+j}$. Then G is really the union of two closed walks: the one from v_i to v_{i+j}, and the walk from v_1 to v_i joined by that from v_{i+j} to v_1. One of the two must have an odd length, since their sum is, and which is clearly less than n, hence it contains a cycle of odd length.   ▽

Theorem 5.12. Let A denote the adjacency matrix of a graph G with vertices $v_1, v_2, \ldots, v_n$. The number of walks of length k from v_i to v_j is then given by $[A^k]_{ij}$.

Proof. It is trivial that $[A]_{ij} = 1$ if and only if there is a walk of length one from v_i to v_j. We proceed by induction. Since a walk of length $k + 1$ from v_i to v_j consists of a walk of length k from v_i to an intermediate vertex v_m, which must be adjacent to v_k, then the total number of such walks is just

$$\sum_{m=1}^{n} [A^k]_{im} [A]_{mj} = [A^{k+1}]_{ij}$$

and the induction is complete.   ▽

   In particular, a closed walk of length 3 is a triangle. Since the vertices of a labeled triangle can be permuted in 6 different ways, we conclude that a labeled graph given by its adjacency matrix A contains

$$\frac{1}{6} \sum_{i \geq 1} [A^3]_{ii}$$

triangles. In matrix algebra, the sum of diagonal entries is called the *trace*. Thus, one-sixth of the trace of A^3 counts the number of triangles.

Question. How many triangles do you see in the graph given in (5.3)?

Exercise 5.35. Find the number of triangles contained in the graphs K_4, $K_{2,2}$, $K_{2,3}$, and K_5.

Question. Can you find an easy formula for the number of triangles contained in K_n?

5.3.1 Distances

Definition. The *distance* between two vertices a and b, denoted by $d(a, b)$, is the length of the shortest walk from a to b, if it exists, or else let $d(a, b) = \infty$.

For example, if a and b are adjacent, then $d(a, b) = 1$. Also, by definition, we have $d(a, a) = 0$. Note that the shortest walk is necessarily a path.

Definition. The *diameter* of a graph G, denoted by $d(G)$, is the largest possible distance between two vertices in G.

Hence, for example, $d(G) = 1$ if and only if G is complete. Note also that $d(G) = \infty$ if and only if G is disconnected.

Question. What is the diameter of the Peterson graph?

Test 5.36. Which one of these four graphs has the largest diameter?

a) K_{99}
b) $K_{99,99}$
c) P_{99}
d) C_{99}

Exercise 5.37. Find the diameters of the graphs K_n, $K_{m,n}$, P_n, and C_n.

Theorem 5.13. If $d(G) \geq 3$ then $d(\overline{G}) \leq 3$.

Proof. Assume $d(G) \geq 3$ and take two vertices v and w. We will show that $d(v, w) \leq 3$ in $\overline{G}$.
  We know there are a and b for which $d(a, b) \geq 3$ in G. It is then false that both $av, vb \in E_G$ or both $aw, wb \in E_G$. Hence, $E_{\overline{G}}$ must contain either av and aw, in which case $d(v, w) \leq 2$ in $\overline{G}$; or vb and wb, again we are done; or av and wb; or else vb and aw. In these last two cases, with the fact that $ab \in E_{\overline{G}}$, we would have $d(v, w) \leq 3$ in $\overline{G}$. $\triangledown$

Exercise* 5.38. If G is self-complementary, prove that $d(G) = 2$ or 3.

Definition. Suppose that $V_G = \{v_1, v_2, \ldots, v_n\}$. The *distance matrix* of the graph G is the $n \times n$ matrix D given by $[D]_{ij} = d(v_i, v_j)$.

For example, the distance matrix of P_4, labeled in the standard way, is given by

$$D = \begin{bmatrix} 0 & 1 & 2 & 3 \\ 1 & 0 & 1 & 2 \\ 2 & 1 & 0 & 1 \\ 3 & 2 & 1 & 0 \end{bmatrix}$$

Question. Can non-isomorphic graphs have identical distance matrices?

Exercise 5.39. Describe the distance matrices of the graphs K_n, $K_{m,n}$, P_n, and C_n.

Exercise 5.40. Convert the incidence matrices given in Exercise 5.23 to distance matrices.

Exercise 5.41. Consider again the adjacency matrix A given in Exercise 5.24. Which one of the following is the corresponding distance matrix?

$$
\text{a)} \begin{bmatrix} 0 & 1 & 1 & 1 \\ 1 & 0 & 2 & 1 \\ 1 & 2 & 0 & 2 \\ 1 & 1 & 2 & 0 \end{bmatrix} \quad
\text{b)} \begin{bmatrix} 0 & 1 & 1 & 1 \\ 1 & 0 & 1 & 1 \\ 1 & 1 & 0 & 1 \\ 1 & 1 & 1 & 0 \end{bmatrix} \quad
\text{c)} \begin{bmatrix} 0 & 1 & 1 & 1 \\ 1 & 0 & 2 & 2 \\ 1 & 2 & 0 & 2 \\ 1 & 2 & 2 & 0 \end{bmatrix} \quad
\text{d)} \begin{bmatrix} 0 & 1 & 1 & 1 \\ 1 & 0 & 2 & 1 \\ 1 & 2 & 0 & 3 \\ 1 & 1 & 3 & 0 \end{bmatrix}
$$

With the help of a computer, it is not hard to convert an adjacency matrix to the distance matrix. The following algorithm shows that what it takes is to compute the powers of A.

Algorithm 5.14 (From A to D). Given the $n \times n$ adjacency matrix A, we are able to retrieve the distance matrix D following these steps.

1) Compute the matrices $A, A^2, A^3, \ldots, A^n$.

2) Set $[D]_{ii} = 0$, and for $i \neq j$ let $[D]_{ij} = k$, the least exponent for which $[A^k]_{ij} \neq 0$. If no such k exists, set $[D]_{ij} = \infty$.

Observation. Theorem 5.12 says that $[A^k]_{ij}$ is the number of walks from i to j of length k. The first nonzero k is therefore the distance, and if there is no walk from i to j of length n or less, then there is no such walk. $\triangledown$

Definition. If G is a weighted graph, we redefine the distance $d(a, b)$ to be the least total weight of all possible walks from a to b.

Question. What is the largest possible distance between two vertices in the weighted graph given in (5.4) earlier?

The next algorithm, which we will present without proof, computes distances in a weighted graph. It reminds us of Prim's algorithm for spanning trees, in the way that it repeatedly search for the least next weight within an increasingly bigger neighborhood.

Algorithm 5.15 (Dijkstra). We compute $d(a, b)$ in a weighted graph and find the shortest path (least total weight) from a to b.

1) We denote by S the set of vertices s which have already been labeled by (v_s, W_s), where v_s is a vertex and W_s is an integer. Initially, S contains only the vertex a, which we label $(-, 0)$.

2) For each vertex $x \in N(S)$, say adjacent to $y \in S$ with weight W, calculate the number $W_x = W + W_y$. Choose the least of such numbers, and label the corresponding vertex x, or vertices if not unique, by (y, W_x).

3) Repeat until b is labeled, in which case $d(a, b) = W_b$. Moreover, the shortest path from a to b can be found by backtracking the first components in the labels.

Exercise 5.42. Visit again the weighted graph given in Exercise 5.34 and apply Dijkstra's algorithm to compute the distance from east to west, i.e., from the far-right vertex to the far-left vertex.

5.3.2 Euler Circuits

Definition. An *Euler walk* in a connected graph G is a walk through all the edges of G without repeating any of them. If an Euler walk is closed, we shall call it an *Euler circuit*.

As an example, we show below a graph which has an Euler walk from a to b by following the labeled edges $e_1, e_2, \ldots, e_8$, in this order.

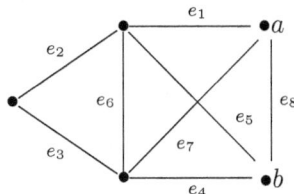

Question. Does the Peterson graph have an Euler walk?

Theorem 5.16. A connected graph has an Euler walk from a to $b \neq a$ if and only if a and b are the only vertices whose degrees are odd. The graph has an Euler circuit if and only if all vertices have even degrees.

Proof. Consider a vertex v with $\deg(v) = d$. At some point during the walk, we will run into v and out via another edge. If $d > 2$, this process will repeat, for as long as there are untrodden edges containing v. This shows the necessity that d be even, unless v is our starting point, or last destination, in which case d may be odd.

To prove sufficiency, assume first that every vertex in a graph G has an even degree. Consider a path P of maximum length from v to w. Since $\deg(w)$ is at least two, w is adjacent to another vertex already contained in P, else we could extend P to a longer path. Hence G contains a cycle

C. Since every vertex in C has an even degree, so does the subgraph whose edges are in $G - C$. Repeating the argument, we see that the edges in G can be thought of as the union of cycles none of which is disjoint from the rest.

We finish off by induction. One cycle is itself an Euler circuit. Assume that the union of n such cycles has an Euler circuit, call it E. With one more cycle C, which meets E at a vertex x, we have an Euler circuit for $E \cup C$ by starting at x, circuiting around E back to x, and cycling around C back to x.

Lastly, if $\deg(a)$ and $\deg(b)$ are the only odd degrees in G, we add one more edge, i.e., ab into G (perhaps making G a multigraph) so that every vertex now has an even degree. We have shown that an Euler circuit exists for this extended graph. Hence, without this extra edge, we could Euler walk from a and terminate at b.   ▽

Exercise* 5.43. Prove that Theorem 5.16 holds for multigraphs as well.

Exercise 5.44. With the help of Theorem 5.16, find an Euler walk or Euler circuit, if any, in each graph given below.

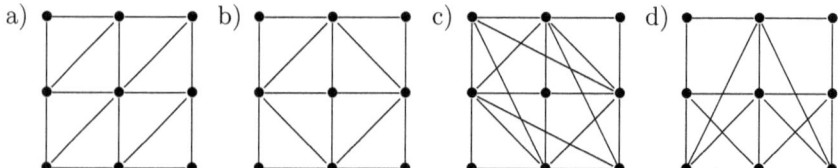

Test 5.45. Which one of these four graphs has an Euler walk but not Euler circuit?

a) K_{99}
b) K_{100}
c) $K_{100,2}$
d) $K_{99,2}$

Exercise 5.46. Determine exactly when each of the graphs K_n, $K_{m,n}$, P_n, and C_n, may have an Euler walk or an Euler circuit.

Amira is a traditional postal carrier who walks through every street within her district of responsibility to deliver the mail. Having studied graph theory, she searches for a closed route which involves the least number of repeated streets or, if possible, none at all.

The *Chinese postman problem* asks for the shortest closed walk going through every edge in a graph, possibly weighted. If any, an Euler circuit would certainly give the optimal solution, or else such a walk would have to repeat one or more edges. The algorithm for solving this problem is next.

Algorithm 5.17 (Chinese Postman Problem). The shortest closed walk covering every edge in the graph, with or without weights, is obtained following these steps.

1) Identify all vertices which have odd degrees. By Euler's theorem, their number is even.

2) Pair up these odd vertices $\{a_1, b_1\}, \{a_2, b_2\}, \ldots, \{a_n, b_n\}$ in such a way that $\sum d(a_i, b_i)$ is the least possible.

3) The shortest walk solution is through all the edges of G, plus the shortest paths from a_i to b_i, where $1 \le i \le n$.

Observation. Any closed walk with no repeated edges will visit each vertex an even number of times, so a vertex with odd degree necessitates a repetition. One would think that these extra walks might well be paths, in order to optimize the length, and that each path should start and end at an odd degree, to make good use of their oddness. These claims, though intuitively acceptable, would need rigorous justifications. ▽

Example. We solve the Chinese postman problem for the weighted graph shown on the left in (5.5). After a brief examination, we find four vertices of odd degree which we label a, b, c, d, in the figure on the right.

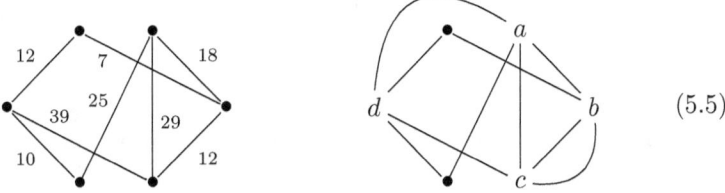

(5.5)

The imaginary edges $\{a, d\}$ and $\{b, c\}$ show the optimal pairing, after exhausting the three possibilities, i.e.,

$$d(a,b) + d(c,d) = (18) + (12 + 7 + 12) = 49$$
$$d(a,c) + d(b,d) = (29) + (7 + 12) = 48$$
$$d(a,d) + d(b,c) = (25 + 10) + (12) = 47$$

Hence, the minimal closed walk is the Euler circuit on the multigraph, whose weight is the overall total 152 plus the extra edge ad, which is really the path of weight $25 + 10$, and bc, of weight 12. These sum to 199 total weight.

Question. What is the minimal length solution to the above example if the graph were unweighted?

Test 5.47. In solving the Chinese postman problem, suppose that there are 6 vertices of odd degree. How many different pairings are possible?

a) 6
b) 12
c) 15
d) 30

Exercise 5.48. Solve the Chinese postman problem for the graph below.

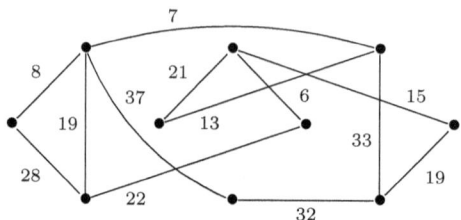

5.3.3   Hamilton Cycles

Definition. A *Hamilton cycle* in a graph G is a cycle $C_n \subseteq G$, where $n = |V_G|$. A *Hamilton graph* is one which contains a Hamilton cycle.

In other words, a Hamilton cycle is a closed walk through all the vertices in G without repeating any of them except, of course, the starting vertex. Note that a Hamilton graph is necessarily connected.

Question. Why is a Hamilton graph never a tree, nor vice versa?

Test 5.49. Which one of these four is *not* a Hamilton graph?

a) K_{99}
b) $K_{99,99}$
c) $K_{99,100}$
d) C_{99}

Exercise 5.50. Determine exactly when each of the graphs K_n, $K_{m,n}$, P_n, and C_n, is a Hamilton graph.

We know no effective algorithms that determine whether a given graph contains a Hamilton cycle, let alone produce one if any. The next theorem states some, rather weak necessary conditions for a Hamilton graph, followed by another theorem on a sufficient condition.

Theorem 5.18. If G is a Hamilton graph, then G contains no leaves, no bridges, and no cut vertices. A *cut vertex* is one that disconnects the graph when removed.

Proof. A Hamilton cycle, or any cycle, is 2-regular. This necessitates every vertex in G to have degree at least two. Now a bridge is the only connection between two components. So a closed walk through both components must cross the bridge twice, hence not a cycle. Similarly, a closed walk through a cut vertex must repeat the vertex, hence not a cycle.   ▽

The converse of the theorem does not hold. The graph given in Exercise 5.48, for example, has no Hamilton cycle. If it did, the vertices of degree two must yield both edges, creating an impossible subcycle shown below.

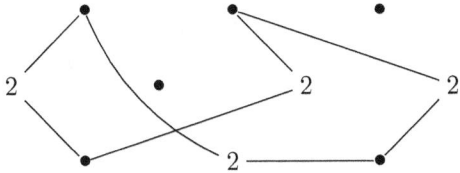

Exercise* 5.51. Prove that the Peterson graph does not contain a Hamilton cycle.

Theorem 5.19. If every vertex in a connected graph G has degree at least $|V|/2$, then G is a Hamilton graph.

Proof. Let $|V| = n$ and P be a path of maximum length in G, given by $v_1v_2, v_2v_3, \ldots, v_{k-1}v_k$. Having maximum length means that the vertex v_1 is not adjacent to any other outside P, and similarly for v_k. Since $\deg(v_1) \geq n/2$, there are at least this many vertices in P adjacent to v_1, and similarly to v_k. By the pigeonhole principle, we can find v_j, with $2 \leq j \leq k$, such that both $v_1v_j, v_{j-1}v_k \in E$. This gives us a cycle from v_1 to $v_{j-1}, v_{j-1}v_k$, backward to v_j, v_jv_1. Then any additional vertex connected to this cycle would contradict the maximality of the length of P, hence $k = n$.   ▽

For example, we may use Theorem 5.19 to show that a complete graph with at least 3 vertices has a Hamilton cycle. But we already know that.

Exercise 5.52. Prove that the complement of the Peterson graph is a Hamilton graph, and so is the complement of C_n, if $n \geq 5$.

Now Elias is a traveling salesman who is planning to visit several major cities for business purposes. Having studied graph theory, Elias searches for a cycling flights through all these cities and back to his hometown, which will cost him the least possible airfares.

Thus the *traveling salesman problem* asks for a Hamilton cycle of least length in a weighted graph. One way to solve the problem is to try out all possible Hamilton cycles and select the one with least total weight, for there is no decisive algorithm that works in all cases.

Example. Let us solve the traveling salesman problem for the following weighted graph.

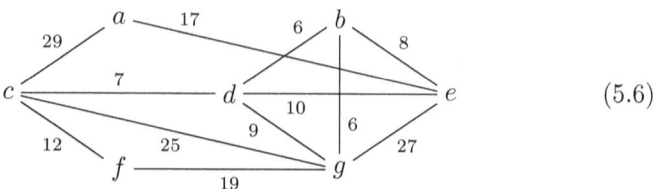

$$(5.6)$$

A cycle takes two edges from each vertex. Due to a and f, of degree two, any Hamilton cycle must contain the path $eacfg$, of weight $17+29+12+19 = 77$. To complete the cycle, we continue with either the path $gbde$, of weight $6+6+10 = 22$, or $gdbe$, of weigth $9+6+8 = 23$. The choice is now obvious: the Hamilton cycle $eacfgbde$, of total weight 99.

Exercise 5.53. Follow the above example and solve the traveling salesman problem with the weighted graph given in (5.5) earlier.

Question. What is the traveling salesman solution to the weighted graph given in (5.4)?

Exercise* 5.54. Find the traveling salesman solution, if any, using the weighted graph given in Exercise 5.34.

Again, there is no generalized technique for solving the traveling salesman problem. And by the way, the shortest closed walk going through every vertex is not always given by a cycle. In the following hypothetical example, taking a walk anywhere is always cheaper via the center vertex!

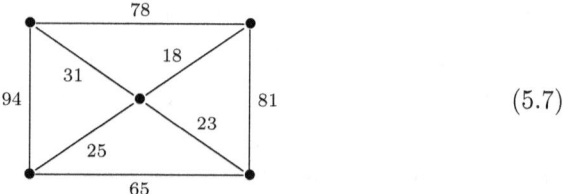

$$(5.7)$$

Exercise 5.55. What will be the least possible total weight for a closed walk through every vertex in the weighted graph given in (5.7)?

5.4   Coloring

One important problem about graphs concerns the idea of vertex coloring. What we mean by vertex coloring is assigning a color to every vertex in such a way that adjacent vertices have different colors. An obvious challenge is then to do this task using as few colors as possible. We discuss first the case where two colors suffice.

5.4.1 Bipartite Graphs

For the next definition, we first introduce the new set notation $X \sqcup Y$, which really stands for the union $X \cup Y$ together with the additional assumption that X and Y are disjoint. We say that $X \sqcup Y$ is the *disjoint union* of X and Y, and that X and Y are the *bipartition subsets* of $X \sqcup Y$.

Definition. A graph G is *bipartite* if $V_G = X \sqcup Y$, such that two vertices are adjacent only if exactly one of them belongs to X (and the other to Y).

Note that if G is disconnected then G is bipartite if and only if each component is bipartite. Hence, in discussing bipartite graphs, we assume that G is a connected graph, unless otherwise stated.

Question. Is the Peterson graph bipartite?

Test 5.56. Which one of these four graphs is *not* bipartite?

a) K_9
b) $K_{8,8}$
c) P_7
d) C_6

The complete bipartite graphs $K_{m,n}$ are but a special subfamily of bipartite graphs, having the property that two vertices are adjacent if and only if they do not belong together in the bipartition subsets.

Theorem 5.20. Let G be any non-trivial graph. The following statements are equivalent one to another.

1) The graph G is bipartite.

2) Two colors are sufficient to color the graph G.

3) There is a complete bipartite graph which contains G.

4) There is no cycle of odd length in G.

Proof. By definition, it is clear that a bipartite graph is a subgraph of some $K_{m,n}$. Moreover, in the case $V = X \sqcup Y$, it suffices to color the vertices in X black and Y white. Hence, if a cycle $v_1v_2, v_2v_3, \ldots, v_nv_1$ is a subgraph, then n is even, or else v_1 and v_n would be of the same color.

To complete the proof, we assume that G contains no cycle of odd length, and we will show G is bipartite. Fix a vertex v and let $X = \{w \in V_G \mid d(v, w)$ is even$\}$ and $Y = \{w \in V_G \mid d(v, w)$ is odd$\}$. It is clear that $V_G = X \sqcup Y$. Moreover, if $a, b \in X$ then there is a walk of even length from v to each of a and b. Then a and b cannot be adjacent, otherwise the closed walk from v to a, then ab, from b back to v, would have an odd length, and that is not allowed by Theorem 5.11. Similarly, if $a, b \in Y$, we see that $ab \notin E$. Thus G is bipartite with these bipartition subsets X and Y. $\triangledown$

Question. Why is a tree always a bipartite graph?

Exercise 5.57. Determine exactly when each of the graphs K_n, $K_{m,n}$, P_n, and C_n, is bipartite.

Exercise* 5.58. Let G be a bipartite graph with n vertices. Prove that $|E_G| \leq n^2/4$, where equality holds if and only if $G \simeq K_{n/2,n/2}$.

Algorithm 5.21 (Bipartite Graph Coloring). We may determine whether G is bipartite, and if so we find the bipartition subsets, as follows.

1) Select an arbitrary vertex and color it black. Let S denote the set of vertices which have already been colored.

2) For each vertex in $N(S)$ color it white or black, such that adjacent vertices have distinct colors. If this is not possible, then G is not bipartite.

3) Repeat until $|S| = |V_G|$. In the end, the vertices are bipartitioned according to their color.

Observation. If G is bipartite, two colors suffice. The color of a vertex is then completely determined by the color of its adjacent vertex, regardless where we start. If G is not bipartite, such procedure is doomed to fail. $\triangledown$

Example. We apply this coloring algorithm to the graph shown on the left. An initial vertex b is chosen black, and we successfully complete the coloring of all seven vertices with black and white. The redrawing on the right shows more apparently that the graph is indeed bipartite.

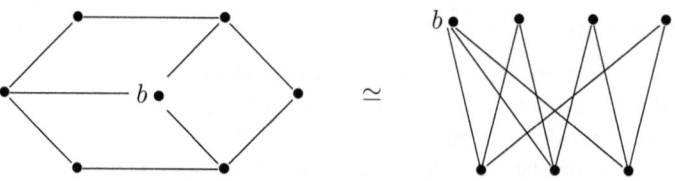

Exercise 5.59. Apply the coloring algorithm to see if the given graph below is bipartite.

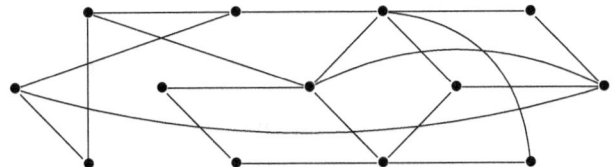

Suppose that Elias runs an online dating service. In terms of graph theory, the bipartition subsets are men and women who are registered members of this service. Agreeably, a match is when a man and a woman have a certain number of common interests in different categories. Knowing that it is not ethical to match two men to the same woman, or vice versa, Elias consults his graph theory text to find the maximum set of matches for his collection of data.

Definition. A *matching* in a bipartite graph G is a subgraph $M \subseteq G$ in which the edges are mutually disjoint. If $V_G = X \sqcup Y$, then a matching M is called *complete* for X if M contains X. And if, in addition, $|X| = |Y|$ then any complete matching (for X or Y) is a *perfect matching*.

Note that if an edge ab is treated like a relation from X to Y, then a matching is just like a one-to-one function, which is complete when the domain is all of X. A perfect matching is therefore like a bijection, which requires that $|X| = |Y|$ and in which case its inverse is also a bijection, from Y to X.

Suppose that G is bipartite with bipartition $V_G = X \sqcup Y$ and a matching M. Let the path $P = \{v_1 v_2, v_2 v_3, \ldots, v_{2n-1} v_{2n}\}$ be such that the edges $v_k v_{k+1} \in M$ if and only if k is even. If, in addition, both $v_1, v_{2n} \notin M$, then replacing the edges in $M \cap P$ by those in $P - M$ will produce a new matching M' which contains one more vertex from each X and Y. Such "alternating" path is the key to proving the next theorem.

Theorem 5.22 (Hall's Theorem). Let G be bipartite with $V_G = X \sqcup Y$. Then G has a complete matching for X if and only if $|S| \leq |N(S)|$ for every subset $S \subseteq X$.

Proof. Necessity is obvious for if $S \subseteq X$ and G has a complete matching M, then in M, every vertex in S is adjacent to its own vertex in Y, thus $|N(S)| \geq |S|$. Now assume, for sufficiency, $|S| \leq |N(S)|$ for every subset $S \subseteq X$. Assume also, by induction, that we have a matching M contaning all of X except one vertex $v \in X$. We will produce an alternating path from v, with respect to M, which will complete this matching for X.

Since $N(\{v\})$ is not empty, we can find $w_1 \in Y$ which is adjacent to v. If $w_1 \notin M$ then vw_1 is such path, we are done. Else, there is an edge $v_1 w_1 \in M$. Since $|N(\{v, v_1\})| \geq 2$, we have w_2, adjacent to either v or v_1. If $w_2 \notin M$, again we have an alternating path from v to w_2, so assume there is another edge $v_2 w_2 \in M$. Continuing in this way, seeing that $|N(\{v, v_1, v_2, \ldots v_k\})| > k$ in each step, we will exhaust the vertices in $Y \cap M$, forcing a vertex $w \notin M$ to which there is an alternating path from v. $\triangledown$

For example, in the bipartite graph given in (5.8) below, a complete matching is not possible since we have $|S| > |N(S)|$ for the set $S = \{a, c, d\}$.

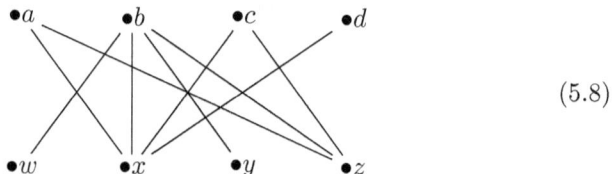

$$\text{(5.8)}$$

Theorem 5.23. Suppose that G is bipartite with $V_G = X \sqcup Y$. If G is regular, then $|X| = |Y|$ and G has a perfect matching.

Proof. Assume that $|X| = n$ and that G is d-regular. Then G has exactly dn edges. Similarly, if $|Y| = m$ then $|E| = dm$. Hence $m = n$. Furthermore, any set S of k vertices in X corresponds to dk edges, hence to k vertices in Y. We have $|N(S)| = |S|$ and a perfect matching by Hall's theorem. $\triangledown$

Exercise 5.60. Draw all regular bipartite graphs with up to 6 vertices.

Exercise* 5.61. Prove that any 2-regular bipartite graph is a cycle of even length, and conversely.

5.4.2   Chromatic Numbers

Definition. The *chromatic number* $\chi(G)$ of a graph G is the least number of colors needed to color the vertices of G, such that adjacent vertices have distinct colors. For example, we have seen that $\chi(G) = 2$ if and only if G is a bipartite graph.

Note that if $H \subseteq G$, then $\chi(H) \leq \chi(G)$. In particular, if G is disconnected then $\chi(G)$ is simply the largest chromatic number among the components of G, so we may well assume henceforth that G is connected where $\chi(G)$ is concerned.

Question. What is the chromatic number of the Peterson graph?

Test 5.62. Which one of these graphs has the largest chromatic number?
a) P_{99}
b) C_{99}
c) P_{100}
d) C_{100}

Exercise 5.63. Determine the chromatic numbers of the graphs K_n, $K_{m,n}$, P_n, and C_n.

Theorem 5.24. Let $\chi(G) = k$. Then the graph G has a vertex v such that $\deg(v) \geq k - 1$, and at least k such vertices.

Proof. Note that removing an edge off G will result in $\chi(G) \leq k$. Let G' be the subgraph of G obtained upon removing as many edges as possible while maintaining $\chi(G') = k$. Hence, removing an edge off G' will make $\chi(G') \leq k - 1$.

It is clear that G' has at least k vertices. We claim that $\deg(v) \geq k - 1$ holds in G', hence in G as well, for each vertex $v \in V_{G'}$. If this were not so, let $\deg(v) \leq k - 2$ in G'. Not counting v in, G' can be colored with $k - 1$ colors or less. So if v has only $k - 2$ adjacent vertices in G', one of the $k - 1$ colors can be assigned to v and we get $\chi(G') \leq k - 1$, a contradiction. $\triangledown$

As a consequence of the preceding theorem, we have a bound for the chromatic number with respect to the degrees of vertices, i.e.,

$$\chi(G) \leq \Delta(G) + 1$$

where $\Delta(G)$ is the largest degree of a vertex in G.

Exercise 5.64. Find two families among the special graphs K_n, $K_{m,n}$, P_n, and C_n, for which $\chi(G) = \Delta(G) + 1$. It can be proved that these are the *only* two classes for which equality holds.

Amira manages the schedule of baseball practice for the Ammon Little League. Practice sessions are to be held at various schools spread throughout town. To avoid family conflict, if there are siblings in two different teams, the two teams should be assigned a different practice day. Wishing to minimize the number of practice days in a week, Amira realizes that the challenge is really to find the chromatic number for her teams.

There are algorithms which generate a coloring solution for G using $\Delta(G)$ colors or less, but there is none so far which can effectively compute $\chi(G)$.

Exercise 5.65. Find the chromatic number for each given graph.
a) The graph given in (5.6).
b) The graph given in (5.5).
c) The graph given in Exercise 5.48.
d) The graph given in Exercise 5.34.

5.4.3   Planar Graphs

Definition. A graph is *planar* if it can be drawn in the plane such that no edges are crossing each other. This particular drawing of a planar graph is called a *plane graph*.

Note that if G is disconnected, then G is planar if and only if each component is planar, hence we may well assume that henceforth G is connected where planarity is concerned.

Question. What is the difference between a planar graph and a plane graph?

For example, although the usual drawing of K_4 involves two edges crossing each other, we may in fact redraw the graph in a way that shows K_4 is planar. We illustrate two possible ways of drawing K_4 as follows.

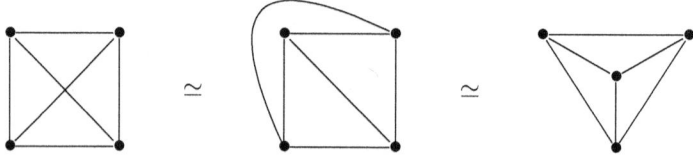

However, you may try in vain to do the same with the graph $K_{3,3}$, for instance, because $K_{3,3}$ is not a planar graph. Since a subgraph of a planar graph is obviously planar, we see that $K_{m,n}$ is not planar if both $m, n \geq 3$.

Question. Can you *prove* that $K_{3,3}$ is not planar?

Exercise 5.66. Prove that $K_{2,n}$ is planar for all $n \geq 1$.

A plane graph partitions the plane into subsets which are called *regions*. Thus, regions refer to the bounded areas interior to the plane graph, plus one unbounded exterior. For example, the plane graph of K_4 has four regions.

Theorem 5.25 (Euler's Formula). Suppose that a connected plane graph has v vertices, e edges, and r regions. Then $v + r = e + 2$.

Proof. If the graph is a tree, then $e = v - 1$ and $r = 1$ since there is no bounded region. Hence, $v + r - e = 2$ as desired. If not a tree, then the graph can be obtained by adding edges to its spanning tree. Each new edge would add a bounded region, increasing the value of r, as well as e, by one. Thus the equality $v + r = e + 2$ is preserved.   ▽

Exercise* 5.67. Prove that Euler's formula holds for connected plane multigraphs as well.

Theorem 5.26. Let G be a planar graph with $n \geq 3$ vertices. The following results are consequences of Euler's formula.

1) The number of edges in G is at most $3n - 6$.

2) If $n \geq 5$ then G is not a complete graph.

3) There is a vertex in G of degree 5 or less.

4) If G contains no triangles, then G has at most $2n - 4$ edges.

Proof. In the drawing of a plane graph, we note that the number of edges is maximized when every region is the interior of a triangle. (With higher polygons, a diagonal edge can be added while keeping G planar.) Since each edge borders two regions, we have $2|E| = 3r$, where r is the number of regions. Substitute $r = 2|E|/3$ in Euler's formula to get $|E| = 3n - 6$, proving (1).

   In particular, K_5 has 10 edges, exceeding the maximum number of $3 \times 5 - 6 = 9$. Hence, K_5 is not planar and neither is K_n for all $n > 5$. Similarly, for (3), if every vertex has degree 6, then by Euler's formula, $|E| \geq 6n/2 = 3n$, violating (1). Finally for (4), simply replace the relation $2|E| = 3r$ above by $2|E| = 4r$ to get the desired result.   ▽

Exercise 5.68. Use Theorem 5.26 to prove that $K_{3,3}$ is not planar.

Question. Can Theorem 5.26 be applied to show that the Peterson graph is not planar?

Exercise* 5.69. Prove that the complement of the Peterson graph is not planar, and neither is the complement of C_n, if $n \geq 8$.

Definition. Let G be a graph. If we replace any edge $ab \in E_G$ by the path $\{av, vb\}$, where v is a new vertex, then the resulting graph is said to be *homeomorphic* to G. More generally, two graphs are *homeomorphic* to each other if one can be obtained from the other by iterating a finite number of replacements in this manner.

   For example, we sketch below how to obtain C_6 by applying the procedure twice to C_4. In this way, it is not hard to see that any two cycles are homeomorphic.

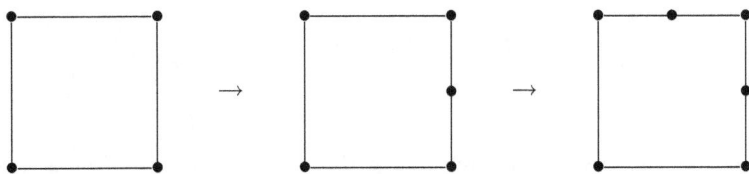

Test 5.70. Which one of these four graphs is *not* homeomorphic to any other one?
a) K_3
b) K_4
c) $K_{2,2}$
d) C_9

Exercise* 5.71. Show that every graph G is homeomorphic to a bipartite graph, by replacing *every* edge in G by a path of length two.

The next major theorem, due to Kuratowski, reduces the question of planarity to a search for either $K_{3,3}$ or K_5 subgraph. The lengthy proof is omitted, but we shall illustrate how one might apply the theorem to show the non-planarity of the Peterson graph.

Theorem 5.27. A graph is planar if and only if it contains no subgraph homeomorphic to K_5 or $K_{3,3}$.

Example. Prove that the Peterson graph is not planar.

Solution. We find a subgraph of the Peterson graph which, after a little modification in the way it is drawn, is shown to be homeomorphic to $K_{3,3}$.

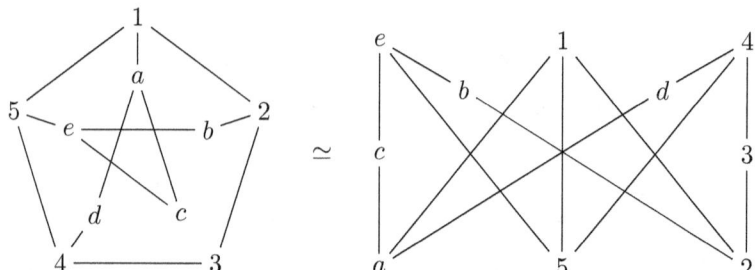

Definition. The *dual graph* G' of a plane graph G is the graph whose vertices are the interior regions of G, and where $rs \in E_{G'}$ if and only if the regions r and s are bounded by a common edge in G.

Below is the illustration of a plane graph G with five interior regions, together with its dual graph G'.

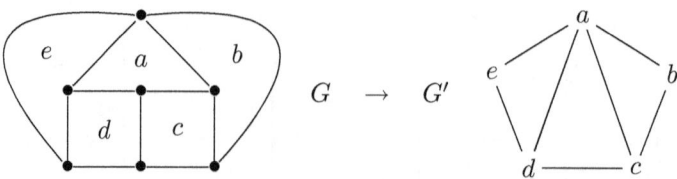

As another example, one may observe that the dual graph of K_4 is K_3.

Test 5.72. Which one of these four is the dual graph of $K_{2,9}$?
a) K_9
b) $K_{8,1}$
c) P_8
d) C_9

Theorem 5.28. The dual graph of any plane graph is planar.

Proof. For each interior region r of the plane graph G, put the vertex r for the dual graph G' somewhere in the middle. To complete the proof, simply observe that we may draw non-intersecting paths from the vertex r to each edge boundary of the region r. $\triangledown$

A plane graph somewhat looks like a world map, in which the interior regions are countries. We say two countries are neighbors when they share a common edge boundary. (Having a common vertex does not make a neighbor.) In fact, map coloring was an early motivation for planar graphs.

Definition. The *chromatic number* of a map is the least number of colors needed to color the countries such that neighbors have distinct colors. In other words, the chromatic number of a map G equals $\chi(G')$.

Exercise* 5.73. Draw the dual graph of the modern day map consisting of the nine countries: Egypt, Iraq, Israel, Jordan, Kuwait, Lebanon, Palestine, Saudi Arabia, and Syria. Then determine the chromatic number.

It was conjectured in the mid nineteenth century that four colors suffice for any map. The four-color theorem remained unproved for over a hundred years, until Appel and Haken gave a complete proof in 1977, involving massive case-by-case figures and extensive computer generated results.

Theorem 5.29 (The Four-Color Theorem). If G is planar then $\chi(G) \leq 4$. Equivalently, the chromatic number of any map is at most four.

Exercise 5.74. Draw a map whose chromatic number equals four.

A version of Appel and Haken's proof which includes sufficient details, spans well over 700 pages. We will compromise, in concluding this chapter, with a shorter and nice partial result, first given by Heawood in 1890.

Theorem 5.30. If G is planar then $\chi(G) \leq 5$.

Proof. We define the *color-degree* of a vertex to be the number of distinct colors of vertices which are adjacent to it.

If G has 5 vertices or less, then there is nothing to prove. We proceed by induction, assuming that G has n vertices and that the claim is true for any planar graph with less vertices than n. By Theorem 5.26, G has a vertex v with $\deg(v) \leq 5$. If we ignore v and its associated edges, the resulting subgraph can then be colored with at most 5 colors. Hence, if the color-degree of v is 4 or less, then we have no challenge assigning v to a fifth color to complete the coloring of G with 5 colors or less.

The last case to consider is when $\deg(v) = 5$ with all five adjacent vertices, v_1, v_2, v_3, v_4, v_5, having distinct colors c_i. Without loss of generality, we have labeled these vertices such that v_3 is interior of the region bounded by the rays vv_1 and vv_2, while v_4 lies exterior of it.

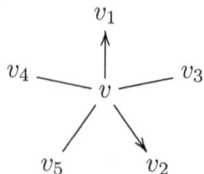

Consider the subgraph $G_{1,2}$ of G consisting of all vertices which have been colored either c_1 or c_2, with their associated edges. Note that $v_1, v_2 \in G_{1,2}$. Suppose first that v_1 and v_2 belong to different components in $G_{1,2}$. In the component containing v_1, we swap c_1 and c_2. Doing so does not violate the rules of vertex coloring, but it does decrease the color-degree of v to 4 as v_1 and v_2 now have the same color, i.e., c_2.

But if v_1 and v_2 belong to the same component, then we have a path from v_1 to v_2 and, combined with v_2v, vv_1, a cycle. Since G is planar, this cycle must enclose either v_3 or v_4, but not both. Again, since G is planar, v_3 and v_4 then belong to different components in the subgraph $G_{3,4}$, defined in a similar way. This time, we decrease the color-degree of v by swapping c_3 and c_4 in the component containing v_3, and the proof is complete. $\triangledown$

Books to Read

1. D. L. Applegate, R. E. Bixby, V. Chvátal, and W. J. Cook, *The Traveling Salesman Problem: A Computational Study*, Princeton University Press 2007.

2. G. Chartrand, *Introductory Graph Theory*, Dover Publications 1984.

3. R. Fritsch and G. Fritsch, *The Four-Color Theorem: History, Topological Foundations, and Idea of Proof*, Springer 1998.

4. D. A. Marcus, *Graph Theory: A Problem Oriented Approach*, Mathematical Association of America 2008.

Answers

Chapter 1

1. (a) 111 (b) 119 (c) 257 (d) 888
2. (c)
3. (a) 1100011 (b) 10000001 (c) 1111100111 (d) 101010101010
4. (a) 170 (b) 3311 (c) 11181 (d) 65793
5. (d)
6. (a) $3E7$ (b) 2711 (c) $181CD$ (d) $7FACE$
7. (a) 511 (b) 668 (c) 3010 (d) 21859
8. (a) 143 (b) 1747 (c) 23421 (d) 300715
9. (a) 1010,1010 (b) 1100,1110,1111
   (c) 10,1011,1010,1101 (d) 1,0000,0001,0000,0001
10. (a) $6F$ (b) $7FF$ (c) 801 (d) $7C39$
11. $2B3A$
12. (a) 157 (b) 2713 (c) 10110100 (d) 1000001000001
13. (a) 5054 (b) 3476702 (c) $BF4$ (d) $1FFFFF$
14. $1B6D$
15. $DONE$
16. Think in base-9 system. Watch out for 13. Answer: 52.
17. (a) 0.25 (b) 0.53125 (c) 0.96875 (d) 0.015625
18. (a) $0.00001_2 = 0.08_{16}$ (b) $0.110001_2 = 0.C4_{16}$
    (c) $0.101_2 = 0.A_{16}$ (d) $0.\overline{01}_2 = 0.\overline{5}_{16}$
19. $\lfloor x \rfloor - 1 \le \lceil x \rceil - 1 \le \lfloor x \rfloor \le x \le \lceil x \rceil \le \lfloor x \rfloor + 1 \le \lceil x \rceil + 1$
20. (a) 3 (b) 35 (c) 0 (d) 99
21. (c)
22. (a)
23. (d)
24. (a) 3 (b) X (c) 0 (d) 7
25. (c)
26. Let me know if there are.
27. (a) 9783314007835 (b) 9789577471345
    (c) 9789622441224 (d) 9789772301546

28. He was lucky.
29. See the back cover.
30. 9783999999999 and 9784000000000, if they count.
31. (a) 25 (b) 12345 (c) 1 (d) 2
32. (a) 15 (b) 3 (c) 1 (d) 26
33. (a) 6325 (b) 2736 (c) 40959 (d) 29000
34. (a) $(5, -8)$ (b) $(3, -22)$ (c) $(418, -1433)$ (d) $(51, -4)$
35. (b)
36. (a) $(95, -152)$ (b) $(21, -154)$ (c) $(4598, -15763)$ (d) No solution.
37. (a) $(95 - 43k, -152 + 69k)$ (b) $(21 - 8k, -154 + 59k)$
    (c) $(4598 - 843k, -15763 + 2890k)$ (d) No solution.
38. Substitute $z = 100 - x - y$ to get the linear equation $28x + 53y = 800$, with unique positive solution $(21, 4, 75)$.
39. 12
40. (a) $2^3 \times 3 \times 37$ (b) $2^5 \times 3^2 \times 5^3$
    (c) $2^2 \times 3^2 \times 7 \times 11 \times 23$ (d) $3 \times 7 \times 11 \times 13 \times 37$
41. (a) prime (b) 29^2 (c) prime (d) 31×37
42. (a) 15 (b) 3 (c) 1 (d) 26
43. (a) 6325 (b) 2736 (c) 40959 (d) 29000
44. (a) 4 (b) 17 (c) 83 (d) 9
45. (b)
46. (a) 1 (b) 2 (c) 4 (d) 9
47. By Theorem 1.11 and Fermat's little theorem, $a^p = a^{p-1}a \equiv a \bmod p$ (mod p), if p does not divide a. If it does, both remainders are zero.
48. $2^{778} \bmod 779 = 605$
49. 341
50. (a) 81 (b) 17×23 (c) 141 (d) 166
51. We already have $m^{eb} \equiv m$ (mod p), which remains valid if p divides m, for both remainders will be zero. With $m^{eb} \equiv m$ (mod q), we see that $m^{eb} - m$ is a multiple of p and q, hence of n. This gives the same result, $s^b \bmod n = m^{eb} \bmod n = m$.
52. (a) No/No (b) Yes/Yes (c) No/No (d) No/Yes
53. (a) $7 \times 13 \times 19$ (b) $5 \times 17 \times 29$ (c) $7 \times 23 \times 41$ (d) $7 \times 19 \times 67$
54. 13 and 73
55. If n is composite with distinct prime factors, one of them is an odd p. Then $p - 1$ is even and divides $n - 1$. So $n - 1$ is even and n is odd.
56. All four.
57. (a) No (b) No (c) Yes (d) Yes
58. In the sequence, each term is the square of the preceding term. So if there is any 1, they stay 1 down to the last term: $a^{n-1} \bmod n = 1$.

Chapter 2

1. (a) T (b) T (c) F (d) F
2. (a) T T T F (b) F T T T (c) T T T T (d) F F F F
3. (a) If $4 + 5 = 9$, then 2 is odd. (F)
   (b) If 2 is odd, then $4 + 5 = 9$. (T)
   (c) If $4 + 5 \neq 9$, then 2 is odd. (T)
   (d) If 2 is even, then $4 + 5 \neq 9$. (F)
4. (a) T F T T (b) T T T T (c) T F T F T F T T (d) T T T T T F F F
5. (a) $(p \wedge q) \to r$ (b) $r \leftrightarrow \neg q$ (c) $p \oplus r$ (d) $(\neg p \wedge \neg q) \to \neg r$
6. (a) F F F F (b) T T T T (c) T F F T F T T F (d) T F F T T F F F
7. (d)
8. (a)
9. (a) tautology (b) contingency (c) contradiction (d) tautology
10. (a) $(\neg p \wedge (p \leftrightarrow q)) \to \neg q$; valid
    (b) $((p \to q) \wedge (\neg p \wedge \neg q)) \to (q \to p)$; valid
    (c) $((p \to \neg q) \wedge (q \to r)) \to (\neg r \to (q \vee \neg p))$; invalid
    (d) $((p \oplus q) \wedge p \wedge (\neg q \to \neg r)) \to (\neg q \wedge \neg r)$; valid
11. (b)
12. (a) F F T F (b) T F F T (c) T T T F F F F F (d) T F T T T T T T
13. F T T T
14. (c)
15. (a) Tomorrow is Sunday, if today is not Friday.
    (b) When x^2 is an integer, so is x.
    (c) A prime number is never even.
    (d) He who is poor cannot be a mathematician.
16. (b)
17. (a) Tomorrow is not Sunday, if today is Friday.
    (b) When x^2 is not an integer, neither is x.
    (c) A composite is always an even number.
    (d) Every wealthy person is a mathematician.
18. (a) 5a, 4c, 4a (b) 5a, 1b, 5a (c) 5a, 3b, 5a (d) 5c, 5b, 5a, 4b, 4c, 4a
19. (a) F (b) T (c) F (d) T
20. (a) $(p \vee \neg q) \wedge (p \vee q) \equiv (p \wedge q) \vee (p \wedge \neg q)$
    (b) $p \vee \neg q \equiv (p \wedge q) \vee (p \wedge \neg q) \vee (\neg p \wedge \neg q)$
    (c) $(\neg p \vee \neg q \vee r) \wedge (p \vee \neg q \vee r) \wedge (p \vee q \vee r) \equiv$
    $(p \wedge q \wedge r) \vee (p \wedge \neg q \wedge r) \vee (p \wedge \neg q \wedge \neg r) \vee (\neg p \wedge q \wedge r) \vee (\neg p \wedge \neg q \wedge r)$
    (d) $(\neg p \vee \neg q \vee \neg r) \wedge (\neg p \vee \neg q \vee r) \wedge (\neg p \vee q \vee r) \wedge (p \neg q \vee \neg r) \wedge$
    $(p \vee \neg q \vee r) \wedge (p \vee q \vee \neg r) \wedge (p \vee q \vee r) \equiv p \wedge \neg q \wedge r$
21. (d)
22. (a) odd numbers (b) even numbers (c) rationals (d) irrationals
23. (d)
24. (a) $\{5, 7\}$ (b) $\{0, 2, 5, 7\}$ (c) $\{5, 7\}$ (d) $\{5, 7\}$
25. (a) A (b) A (c) ϕ (d) ϕ

26. (a)
27. We are to show that $p \oplus q \equiv (p \vee q) \wedge \neg(p \wedge q)$.
28. (a) A (b) $B - A$ (c) $A \oplus B$ (d) $A \cap B$
29. (c)
30. (c)
31. We see this from the Venn diagram or by the use of logical argument, since the proposition $((p \to q) \wedge (q \to r)) \to (p \to r)$ is a tautology, by Theorem 2.1.
32. That $B - A \subseteq A \oplus B$ is clear. Now let $x \in A \oplus B$. Then $x \in A - B$ or $x \in B - A$. But $A - B = \phi$ since $A \subseteq B$, so we must have $x \in B - A$, proving that $A \oplus B \subseteq B - A$.
33. (a) $\{\phi, \{a\}, \{b\}, \{c\}, \{a, b\}, \{a, c\}, \{b, c\}, \{a, b, c\}\}$
    (b) $\{\phi, \{2\}, \{3\}, \{4\}, \{5\}, \{2, 3\}, \{2, 4\}, \{2, 5\}, \{3, 4\}, \{3, 5\}, \{4, 5\}, \{2, 3, 4\}, \{2, 3, 5\}, \{2, 4, 5\}, \{3, 4, 5\}, \{2, 3, 4, 5\}\}$
    (c) $\{\phi\}$
    (d) $\{\phi, \{x\}, \{\{7\}\}, \{x, \{7\}\}\}$
34. (a) 64 (b) 16 (c) 2 (d) 4
35. (d)
36. No, see this example in Section 2.3.3.
37. (a) T (b) T (c) F (d) F
38. (a) $p : x$ is odd $\to x = 2n + 1 \to x^3 = (2n + 1)^3 \to x^3 = 8n^3 + 12n^2 + 6n + 1 \to x^3 = 2(4n^3 + 6n^2 + 3n) + 1 \to x^3 = 2m + 1$, where $m = 4n^3 + 6n^2 + 3n$ is an integer $\to x^3$ is odd: q.
    (b) $x = 2n \to x^2 - 4x + 2 = 4n^2 - 8n + 2 = 2(2n^2 - 4n + 1) \to x^2 - 4x + 2 = 2m \to x^2 - 4x + 2$ is even.
    (c) $p : x$ and y are odd $\to x = 2n + 1$ and $y = 2m + 1 \to x + y = 2(n + m + 1) \to x + y$ is even: q.
    (d) $x = a/b$ and $y = c/d$, where a, b, c, d are integers and $b, d \neq 0 \to x + y = (ad + bc)/bd \to x + y$ is a rational number.
39. (a) If even, $x = 2n$, then $x^3 = 8n^3$ is also even.
    (b) If $x - 3 = \frac{a}{b}$ then $x^2 - 3 = (3 + \frac{a}{b})^2 - 3 = \frac{6b^2 + 6ab + a^2}{b^2}$, which is again rational.
    (c) If both x and y were even, then $x + y$ would be even.
    (d) Because the product of two odd numbers is odd.
40. $x = 2n + 1 \to x^2 - 1 = 4n^2 + 4n = 4n(n + 1)$. One of the consecutive numbers, n or $n + 1$, must be even, hence $x^2 - 1$ is divisible by 8.
41. Suppose $p \bmod 3 = 1$. Then $p = 3n + 1$ for some integer $n > 1$. We know that p is odd and will prove that n is even using contrapositive: because if $n = 2m + 1$ then $p = 3(2m + 1) + 1 = 2(3m + 2)$, and p would be even. Hence $n = 2m$, and so $p = 6m + 1$ and $p \bmod 6 = 1$.
42. (a) If even, $x = 2n \to x^2 + x = 2(2n^2 + n) \to x^2 + x$ is even. If odd, $x = 2n + 1 \to x^2 + x = 2(2n^2 + 3n + 1) \to x^2 + x$ is even.
    (b) If odd, $x = 2n + 1 \to x^2 + 2 = 4n^2 + 4n + 3 \to (x^2 + 2) \bmod 4 =$

$3 \neq 0 \rightarrow x^2 + 2$ is not a multiple of 4. If even, $x = 2n \rightarrow x^2 + 2 = 4n^2 + 2 \rightarrow (x^2 + 2) \bmod 4 = 2 \neq 0 \rightarrow x^2 + 2$ is not a multiple of 4.

(c) If $x = 3n$ then $x^3 - x = 3(9n^3 - n)$. If $x = 3n + 1$ then $x^3 - x = 3(9n^3 + 9n^2 + 2n)$. If $x = 3n + 2$ then $x^3 - x = 3(9n^3 + 18n^2 + 9n + 2)$. In each case, $x^3 - x$ is a multiple of 3.

(d) If $x \geq 0$ and $y \geq 0$, then $xy \geq 0$ and $|xy| = xy = |x| \, |y|$. If $x < 0$ and $y \geq 0$, then $xy \leq 0$ and $|xy| = -xy = (-x)y = |x| \, |y|$. The other two cases are quite similar.

43. There are six cases, $0 \leq p \bmod 6 \leq 5$. Since p must be odd, only three cases apply, i.e., $p = 6n + 1$, $p = 6n + 3$, and $p = 6n + 5$, but the case $6n + 3 = 3(2n + 1)$ applies only to composites. Lastly, note that $6n + 5 = 6m - 1$ with another integer $m = n + 1$.

44. (a) We have shown that if x is even then $x^2 - 4x + 2$ is even. Now show that if x is odd then $x^2 - 4x + 2$ is also odd.

(b) The product of two odd numbers is odd. Conversely, if one of the two numbers is even, then their product is also divisible by 2.

(c) If x or y is a multiple of 5, clearly xy is too. If xy is divisible by 5, then so is x or y, according to Theorem 1.7.

(d) If x is even and y odd, we can show that $x + y$ is odd. If both x and y are even, or both odd, show that $x + y$ is even.

45. Recall the identity $A \oplus B = (A - B) \cup (B - A)$. Note that $A - B$ and $B - A$ are disjoint. Hence, $A \oplus B = B - A$ if and only if $A - B = \phi$, i.e., if and only if $A \subseteq B$.

46. Write $m = \prod p_i^{f_i}$ and $n = \prod p_i^{e_i}$ as in Theorem 1.9. If $m = n$ then $e_i = f_i$ for all, hence $\min\{e_i, f_i\} = \max\{e_i, f_i\}$ and $\gcd(m, n) = \mathrm{lcm}(m, n)$. If $m \neq n$ then $e_i \neq f_i$ for at least one, where $\min\{e_i, f_i\} \neq \max\{e_i, f_i\}$ and $\gcd(m, n) \neq \mathrm{lcm}(m, n)$.

47. (a) Suppose $\sqrt[3]{3} = a/b$, where $\gcd(a, b) = 1$. Then $a^3 = 3b^3$. Since 3 divides a^3, then 3 divides a by Theorem 1.7. We write $a = 3n$, hence $9n^3 = b^3$. Similarly, 3 divides b, contradicting $\gcd(a, b) = 1$.

(b) Suppose $\sqrt[n]{p} = a/b$ with $\gcd(a, b) = 1$. Then $a^n = pb^n$. By Theorem 1.7, p divides a. Let $a = pm$ so that $p^{n-1}m^n = b^n$. But then p divides b, a contradiction.

(c) Suppose $\log_{10} 2 = a/b$. Then $10^a = 2^b$. Why does this contradict the fundamental theorem of arithmetic?

(d) Let x be an irrational number, and let $y = a/b$. Suppose $x + y = c/d$, rational. Then $x = c/d - y = (bc - ad)/bd$, a rational number.

48. Let $a = 2m + 1$ and $b = 2n + 1$. Suppose there is an integer n such that $n^2 = a^2 + b^2 = 4(m^2 + n^2 + m + n) + 2$. Then $n^2 \bmod 4 = 2$. But n^2 is even, hence n is even. Let $n = 2k$. Then $n^2 = 4k^2$ and $n^2 \bmod 4 = 0$, a contradiction.

49. (a) Let $a = 4m + 1$ and $b = 4n + 1$. Then $ab = 4(4mn + m + n) + 1$ and $ab \bmod 4 = 1$.

(b) Being a remainder, $p \bmod 4 = 0, 1, 2,$ or 3. But $p \bmod 4 = 0$ means that p is a multiple of 4, not a prime. Similarly, the case $p \bmod 4 = 2$ implies $p = 4k + 2 = 2(2k + 1)$, a multiple of 2.

(c) Let $n \bmod 4 = 3$ and factor n into primes. If each factor p satisfies $p \bmod 4 = 1$, then by (a) we would have $n \bmod 4 = 1$, a contradiction. Hence by (b) there must be a prime factor p with $p \bmod 4 = 3$.

(d) Suppose only $p_1, \ldots, p_r$ exist. Let $N = 4(p_1 \cdots p_r - 1) + 3$. Since $N \bmod 4 = 3$, by (c) there is one p_i dividing N. The same p_i also divides $N - 4p_1 \cdots p_r = -1$, a contradiction.

50. If not then $n = pq$ and $n - 1 = (p - 1)q + (q - 1)$. Since $p - 1$ divides $n - 1$, this means $p - 1$ divides $q - 1$. Exchanging the roles of p and q, we see that $q - 1$ divides $p - 1$ too. Hence, $p - 1 = q - 1$ and $p = q$. But a Carmichael number has distinct prime factors, a contradiction.

51. (a) T F (b) F F (c) T F (d) T T

52. Represent the union by X and the intersection Y. It is clear $A_1 \subseteq X$. To see $X \subseteq A_1$, let $x \in X$. Then $\exists n \in \mathbb{N} : x \in A_n$. But $A_n \subseteq A_1$, so $x \in A_1$. Similarly, it is clear $0 \in Y \subseteq A_1$. Now if $x > 0$ then $x \notin A_n$ when $n > 1/x$, since $1/n < x$, hence $x \notin Y$. This shows $Y = \{0\}$.

53. (a) T (b) T (c) T (d) T

54. (a) F F F T (b) F T T T (c) T T T T (d) T T F T

55. (b)

56. (a) $p = 29$

(b) $x^2 + 1 = 0$

(c) If $\sqrt{p} = n \in \mathbb{N}$ then $p = n^2$, a composite.

(d) We have $x^2 - 6x + 11 = (x - 3)^2 + 2 \geq 2 \forall x \in \mathbb{R}$ since $(x - 3)^2 \geq 0$.

57. Having proved that $x = \log_2 3$ is irrational, it is easy to show that $2x$ is again irrational. Then $(\sqrt{2})^{2x} = 2^x = 3 \in \mathbb{Z}$.

58. Note that $(-1)^3 = -1$ to see existence. If both $a^3 = -1$ and $b^3 = -1$, then $a^3 = b^3$. It follows that $0 = a^3 - b^3 = (a - b)(a^2 + ab + b^2)$. Either $a = b$ or else $0 = a^2 + ab + b^2 = (a + \frac{b}{2})^2 + \frac{3b^2}{4}$, which implies that $b = 0$ and $a = 0$. Either way, $a = b$.

59. Every $m/n \in \mathbb{Q}$, dividing m, n by their gcd if necessary, can be written with $\gcd(m, n) = 1$. To prove uniqueness, suppose $m/n = a/b$ with $\gcd(a, b) = 1$, hence $bm = an$. Since m divides an and m shares no common factors with n, then $a = mc$ for some $c \in \mathbb{N}$. Similarly, $b = nd$. But $m/n = mc/nd$, so $c = d = 1$ and $(m, n) = (a, b)$.

60. (a) $P(1) : 1 = 2 - 1$, true. $P(n) : 1 + 2 + 4 + \cdots + 2^{n-1} = 2^n - 1 \rightarrow$ $1 + 2 + 4 + \cdots + 2^{n-1} + 2^n = 2^n - 1 + 2^n \rightarrow 1 + 2 + 4 + \cdots + 2^{n-1} + 2^n = 2^{n+1} - 1 : P(n + 1)$.

(b) $P(1) : 1 = \frac{3-1}{2}$, true. $P(n) : 1 + 3 + 9 + \cdots + 3^{n-1} = \frac{3^n - 1}{2} \rightarrow$

$1+3+9+\cdots+3^{n-1}+3^n = \frac{3^n-1}{2}+3^n = \frac{3^n-1+2(3^n)}{2} = \frac{3^{n+1}-1}{2} : P(n+1)$.

(c) $P(1) : 1 = \frac{1\times2\times3}{6}$, true. $P(n) : 1+4+9+\cdots+n^2 = \frac{n(n+1)(2n+1)}{6} \to$
$1+4+9+\cdots+n^2+(n+1)^2 =$
$\frac{n(n+1)(2n+1)}{6}+(n+1)^2 = \frac{(n+1)(n(2n+1)+6(n+1))}{6} =$
$\frac{(n+1)(2n^2+7n+6)}{6} = \frac{(n+1)(n+2)(2n+3)}{6} : P(n+1)$.

(d) $P(1) : 1 = \frac{1\times1\times3}{3}$, true. $P(n) : 1+9+25+\cdots+(2n-1)^2 =$
$\frac{n(2n-1)(2n+1)}{3} \to 1+9+25+\cdots+(2n-1)^2+(2n+1)^2 =$
$\frac{n(2n-1)(2n+1)}{3}+(2n+1)^2 = \frac{(2n+1)(n(2n-1)+3(2n+1))}{3} =$
$\frac{(2n+1)(2n^2+5n+3)}{3} = \frac{(2n+1)(n+1)(2n+3)}{3} : P(n+1)$.

61. (a) It is true $3^7 < 7!$. For $n \geq 7$, assume that $3^n < n!$. Then $3^{n+1} = 3(3^n) < 3n! < (n+1)n! = (n+1)!$.

(b) It is true $3^2 > 1+2^2$. For $n \geq 2$, assume that $3^n > 1+2^n$. Then $3^{n+1} > 2(3^n) > 2(1+2^n) > 1+2^{n+1}$.

(c) We have $5^2 < 2^5$. If $n \geq 5$ and $n^2 < 2^n$, then $(n+1)^2 = n^2+2n+1 < n^2+4n < 2n^2 < 2(2^n) = 2^{n+1}$.

(d) We have $2^2 > 2!$. If $n \geq 2$ and $n^n > n!$, then $(n+1)^{n+1} = (n+1)(n+1)^n > (n+1)n^n > (n+1)n! = (n+1)!$.

62. (c)

63. (a) True for $n = 1$. If $2^{2n}-1 = 3k$, then $2^{2(n+1)}-1 = 2^2(2^{2n})-1 = 4(3k+1)-1 = 3(4k+1)$, again a multiple of 3.

(b) True for $n = 1$. If $n^3+2n = 3k$, then $(n+1)^3+2(n+1) = n^3+3n^2+3n+1+2n+2 = 3(k+n^2+n+1)$, a multiple of 3.

(c) True for $n = 1$. If $n^5-n = 5k$, then $(n+1)^5-(n+1) = 5k+5n^4+10n^3+10n^2+5n$, still divisible by 5.

(d) True for $n = 1$. If $2^{n+2}+3^{2n+1} = 7k$, then $2^{n+3}+3^{2n+3} = 2(2^{n+2})+9(3^{2n+1}) = 9(7k)-7(2^{n+2})$, again divisible by 7.

64. Hint: the case $n = 2$ is given by Theorem 2.6. The second identity can be derived from the first: $\cup\neg A_n = \neg(\neg\cup\neg A_n) = \neg\cap\neg(\neg A_n) = \neg\cap A_n$.

65. Let $P(n)$ denote the statement "n is a product of prime numbers." Then $P(2)$ is trivially true. Assume $P(2), P(3), \ldots, P(n)$ are all true. If $n+1$ is prime, then $P(n+1)$ is also true. If $n+1$ is composite, then $n+1 = ab$ with $2 \leq a,b < n+1$. By assumption, both $P(a)$ and $P(b)$ are true, i.e., a and b are each product of primes. We see that $n+1 = ab$ is also a product of primes, hence $P(n+1)$ is true.

Chapter 3

1. (d)
2. We note $(c,a) \in R^{-1} \circ S^{-1} \leftrightarrow \exists(c,b) \in S^{-1} \wedge (b,a) \in R^{-1} \leftrightarrow \exists(a,b) \in R \wedge (b,c) \in S \leftrightarrow (a,c) \in S \circ R$, thus $R^{-1} \circ S^{-1}$ and $S \circ R$ are inverses.

3. (a) $\{(1,1),(2,2),(2,3),(2,4),(3,3),(4,1)\}$
   (b) $\{(1,1),(2,1),(2,2),(2,3),(2,4),(3,3),(4,1)\}$
   (c) $\{(1,1),(1,4),(2,2),(3,2),(3,3),(4,2)\}$
   (d) $\{(1,1),(2,2),(2,4),(3,3),(4,2),(4,4)\}$

4. We note $(a,b) \in R^n$ if and only if there are n "telescoping" elements $(a,r_1),(r_1,r_2),\ldots,(r_{n-2},r_{n-1}),(r_{n-1},b)$ in R. And $(a,b) \in R^m \circ R^n$ if and only if such, with length $n+m$. The same holds for R^{m+n}.

5. We have $(a,c) \in R \circ A^0 \leftrightarrow \exists (a,b) \in A^0 \wedge (b,c) \in R$. The latter holds if and only if $b = a$ and $(a,c) \in R$, so $R \circ A^0 = R$. Similar for $A^0 \circ R$.

6. (a) symmetric (b) reflexive, anti-symmetric, and transitive
   (c) reflexive, symmetric, transitive (d) anti-symmetric and transitive

7. Any subset of A^0.

8. Many possible examples.

9. These depend on your answers in 3.8.

10.
$$\begin{bmatrix} 0 & 0 & 1 & 0 \\ 1 & 0 & 0 & 1 \\ 0 & 0 & 1 & 0 \\ 0 & 1 & 0 & 0 \end{bmatrix}$$

11. (d)

12. Let $(i,j)_2 = 0$ if and only if $\sum_{k=1}^{n} (i,k) \times (k,j) = 0$.

13. (a) $\{(1,1),(1,2),(1,3),(1,4),(2,1),(2,2),(2,3),(2,4),(3,4)\}$
    (b) $A \times A - \{(1,2),(1,3),(1,4)\}$ (c) $R \cup \{(2,4)\}$ (d) $A \times A$

14. (a)

15. We have proved that $\overline{R}$ is transitive. Conversely, if R is transitive, then the smallest transitive relation which contains R is R itself.

16. (a) $R \cup A^0$ (b) $R \cup R^{-1}$

17. (a) even and odd classes (b) $\{1\},\{2\},\{3\},\{4\}$
    (c) $\{0,9\},\{5,8,11\},\{10\}$ (d) $\{1,9\},\{2,6\},\{3,7,11\},\{12\}$

18. (b)

19. (d)

20. (a) ⋮ 3 | 2 | 1 (b) 36 / 12 18 / 2 3 (c) (d) A ... $\emptyset$

21. The relation is reflexive, which is trivial, transitive (see Exercise 2.31) and anti-symmetric by Theorem 2.7.

22. (c)

23. (a) and (c)

24. (d)

25. Either (a,b) or $(b,a) \in R$ for every $a,b \in A$. Hence $R \cup R^{-1} = A \times A$.

26. (c)

27. (a) $f(\mathbb{R}) = \mathbb{Z}$ (b) $f(\mathbb{R}) = \{0, 1\}$
    (c) $f((-1, \infty) - \{0\}) = \mathbb{R} - \{0\}$ (d) $f(\mathbb{Z} \times \mathbb{Z} - \{(0,0)\}) = \mathbb{N}$
28. (a) one-to-one (b) onto (c) both (d) neither
29. (c)
30. (c)
31. See Theorem 3.10.
32. (a) $g(f(x)) = 4x^2 - 4x + 2$ (b) $g(f(x)) = x$ (c) $g(f(n)) = n$
    (d) $g(f(n)) = 0$ if $n \geq 0$ and $g(f(n)) = 1$ if $n \leq -2$.
33. (c)
34. (a) $(-\infty, 0]$ (b) $[0, 1)$ (c) $[1]_2$ (d) $\emptyset$
35. (b)
36. Reflexive via $\iota\delta_A$ and transitive by Theorem 3.8.
37. (d)
38. If $A \subseteq B$ then $|A| \preceq |B|$. If $|B| \preceq \aleph_0$ then $|A| \preceq \aleph_0$ by transitivity.
39. Let $A = \{a_1, a_2, a_3, \ldots\}$ and $B = \{b_1, b_2, b_3, \ldots\}$, assumed disjoint or
    else replace B by $B - A$. Then $f : A \cup B \to \mathbb{N}$ given by $f(a_n) = 2n$
    and $f(b_n) = 2n + 1$ is clearly an injection.
40. By induction, assuming $A = A_1 \cup A_2 \cup \cdots \cup A_n$ is countable, then
    $A \cup A_{n+1}$ is countable by the previous exercise.
41. Let $f : A \to \mathbb{N}$ and $g : B \to \mathbb{N}$ be two injections. Then the function
    $f \times g : A \times B \to \mathbb{N} \times \mathbb{N}$ given by $f \times g(a, b) = (f(a), g(b))$ is also
    one-to-one. The injection $h(m, n) = 2^m \times 3^n$ can be used again to
    construct the injective composition $h \circ f \times g : A \times B \to \mathbb{N}$.
42. Use the bijection $f(x) = e^x$.
43. Since $a+b+ab+1 = (a+1)(b+1)$, we have $a \star b \neq -1$ whenever $a, b \in G$,
    affirming that $\star$ is a binary operation on G. The identity is $e = 0$. And
    for $a \in G$, note that $a \star -a/(1+a) = (a + a^2 - a - a^2)/(1+b) = 0$, so
    we have inverses. Lastly, associative is true as we can check $(a \star b) \star c =$
    $(a+b+ab) \star c = a+b+ab+c+(a+b+ab)c = a+b+c+bc+(b+c+bc)a =$
    $a \star (b + c + bc) = a \star (b \star c)$.
44. (c)
45. It follows since $(ab)(b^{-1}a^{-1}) = a(bb^{-1})a^{-1} = aa^{-1} = e$.
46. Call the identities e_G of G and e_H of H. The identity of $G \times H$ is then
    (e_G, e_H), and inverse is given by $(g, h)^{-1} = (g^{-1}, h^{-1})$. Associativity
    is inherited from G and H in each component.
47. Necessity is obvious for each one. For sufficiency:
    (a) Since $b = a^{-1}ab = baa^{-1}$, then $ab = ba$.
    (b) Since $b = aa^{-1}b = ba^{-1}a$, then $ab = ba$.
    (c) If $(ab)(ab) = aabb$, then $ba = ab$.
    (d) If $(ab)^{-1} = a^{-1}b^{-1}$, then $ab = (a^{-1}b^{-1})^{-1} = (b^{-1})^{-1}(a^{-1})^{-1} = ba$.
48. We have $ab = a(ab)^2b = a(ab)(ab)b = a^2bab^2 = ba$ for all $a, b \in G$.
49. $\mathbb{Z}_5$ and $\mathbb{Z}_7$.
50. Since $b +_n c \in [b + c]_n$, then $a \times_n (b +_n c) \in [a(b + c)]_n$. And since
    $0 \leq a \times_n (b +_n c) < n$, then $a \times_n (b +_n c) = (ab + ac) \bmod n$.

$\times_{12}$	1	5	7	11
1	1	5	7	11
5	5	1	11	7
7	7	11	1	5
11	11	7	5	1

51. (table above)

52. The pairs $(1,1), (2,6), (3,4), (5,9), (7,8), (10,10)$ are inverses.

53. (c)

54. Matrix multiplication is associative. The identity in $SL(2,\mathbb{R})$ is the identity matrix I, where $\det I = 1$. If $A \in SL(2,\mathbb{R})$ then $\det A^{-1} = \det A$, so $A^{-1} \in SL(2,\mathbb{R})$. Lastly, if A and B have determinant ± 1, so does AB, affirming the binary operation in $SL(2,\mathbb{R})$.

55. Associative is true as in G. Since $e \in H$ and $e \in K$, then $e \in H \cap K$. And if $a \in H \cap K$, then $a^{-1} \in H$ and $a^{-1} \in K$, hence $a^{-1} \in H \cap K$.

56. No. Both $2\mathbb{Z}$ and $3\mathbb{Z}$, the sets of multiples of 2 and 3, are subgroups of $\mathbb{Z}$ under addition. The sum $2 + 3 = 5$ is not even in their union.

57. Let $H = \{n/2 \mid n \in \mathbb{Z}\}$.

58. (a) $5m, 5n \in H \to 5m + (-5n) = 5(m - n) \in H$.
   (b) $\pi^m, \pi^n \in H \to \pi^m \times \pi^{-n} = \pi^{m-n} \in H$.
   (c) If $c^2 - 2d^2 = 1$ then $(a+b\sqrt{2})(c+d\sqrt{2})^{-1} = (ac-2bd)+(bc-ad)\sqrt{2}$, belongs to H since $(ac-2bd)^2 - 2(bc-ad)^2 = (a^2 - 2b^2)(c^2 - 2d^2) = 1$.
   (d) $\det A, \det B \in \{\pm 1\} \to \det(AB^{-1}) = \det A / \det B \in \{\pm 1\}$.

59. Let $a \in H = \{a_1, \ldots, a_n\}$. Then $\{aa_1, \ldots, aa_n\} \subseteq H$. By cancellation law, if $aa_i = aa_j$ then $a_i = a_j$. So this subset is all of H. In particular, there is $aa_i = a$, hence $a_i = e \in H$, and there is $aa_j = e$, hence $a_j = a^{-1} \in H$, passing the subgroup test.

60. If $ax = xa$ then $x^{-1}(ax)x^{-1} = x^{-1}(xa)x^{-1}$ and $x^{-1}a = ax^{-1}$. If moreover $ay = ya$, then $a(xy) = (ax)y = (xa)y = x(ay) = x(ya) = (xy)a$. This shows that $C(a)$ passes the two-step test.

61. If $j, k \geq 0$ then $a^j a^k = aa \cdots a = a^{j+k}$. If both $j, k \leq 0$, then $a^j a^k = (a^{-1})^{-j}(a^{-1})^{-k} = (a^{-1})^{-j-k} = a^{j+k}$ by (1). If only $k \leq 0$, then $a^j a^k = a^j(a^{-1})^{-k}$. This equals $a^{j-(-k)}$ if $j \geq -k$, otherwise $(a^{-1})^{-k-j} = a^{j+k}$. Similarly for $j \leq 0$, as well as for the last claim.

62. (a) 1, 3, 5, 7 (b) 1, 2, 4, 5, 7, 8 (c) 1, 3, 7, 9
   (d) 1, 2, 3, 4, 5, 6, 7, 8, 9, 10

63. (b)

64. (a) 2, 7, 8, 10 (b) none (c) 2, 6, 7, 11 (d) 3, 5

65. Let $G = \{e, a, b\}$. Then $ab = e$ and $a^2 b = a$. In particular, $a^2 \neq e$ and $\{e, a, a^2\} = G$. So a generates G.

66. $\mathbb{Z}_2 \times \mathbb{Z}_3$ is cyclic, $\mathbb{Z}_2 \times \mathbb{Z}_2$ is not.

67. Reflexive: $aa^{-1} = e \in H$. Symmetric: $ab^{-1} \in H \to (ab^{-1})^{-1} = ba^{-1} \in H$. Transitive: $ab^{-1} \in H \wedge bc^{-1} \in H \to ab^{-1}bc^{-1} = ac^{-1} \in H$.

68. (a) $\langle 18 \rangle = \{0, 6, 12, 18\}$, $\langle 18 \rangle 1 = \{1, 7, 13, 19\}$, $\langle 18 \rangle 2 = \{2, 8, 14, 20\}$
    $\langle 18 \rangle 3 = \{3, 9, 15, 21\}$, $\langle 18 \rangle 4 = \{4, 10, 16, 22\}$, $\langle 18 \rangle 5 = \{5, 11, 17, 23\}$.
    (b) $\langle 3 \rangle = \{1, 3, 9\}$, $\langle 3 \rangle 2 = \{2, 5, 6\}$, $\langle 3 \rangle 4 = \{4, 10, 12\}$, $\langle 3 \rangle 7 = \{7, 8, 11\}$.
    (c) $[a]_5 = \{5k + a \mid k \in \mathbb{Z}\}$ for $a = 0, 1, 2, 3, 4$.
    (d) $\{\pm a\}$ for every $a > 0$, rational number.

69. $H \cap K$ is a common subgroup of H and K, hence $|H \cap K|$ is a common divisor of $|H|$ and $|K|$, i.e., $|H \cap K| = 1$.

70. If $a \in G$, the subgroup $\langle a \rangle$ has order a divisor of $|G|$. If $|G|$ is prime, and $a \neq e$, then $|\langle a \rangle| = |G|$, i.e., $\langle a \rangle = G$.

71. 12, 6, 4, 3, 12, 2, 12, 3, 4, 6, 12, 1.

72. (a)

73. $p^n - p^{n-1}$

74. Let $G = \langle a \rangle$ of order m, and $H = \langle b \rangle$ of oreder n. If $(a, b)^k = (e_G, e_H)$, then k is a common multiple of m and n by Theorem 3.31. If $\gcd(m, n) = 1$ then $\mathrm{lcm}(m, n) = mn$. We conclude that in $G \times H$, $|(a, b)| = mn = |G \times H|$, i.e., (a, b) generates $G \times H$.

75. Let $(a, b) \in \mathbb{Z}_m \times \mathbb{Z}_n$. Theorem 3.29 says that $a^m = 0 \in \mathbb{Z}_m$ and $b^n = 0 \in \mathbb{Z}_n$. Hence, $(a, b)^{\mathrm{lcm}(m,n)} = (0, 0) \in \mathbb{Z}_m \times \mathbb{Z}_n$. But $\mathrm{lcm}(m, n) = mn/\gcd(m, n) < mn = |\mathbb{Z}_m \times \mathbb{Z}_n|$ if $\gcd(m, n) > 1$, in which case (a, b) cannot generate $\mathbb{Z}_m \times \mathbb{Z}_n$.

76. (c)

77. Since $\mathbb{Z}_n = \langle 1 \rangle$ and $m = m1$, by Theorem 3.33 $\mathbb{Z}_n = \langle m \rangle$ if and only if $\gcd(m, n) = 1$, i.e., $m \in \mathbb{U}_n$.

78. (b)

79. Since $|\mathbb{U}_n| = \phi(n)$, there are $\phi(\phi(n))$ by Theorem 3.33.

80. (c)

81. Let G be cyclic of order n. By Lagrange's theorem, $|a|$ divides n for each $a \in G$. By Theorem 3.33, there are $\phi(d)$ elements of order d, for each divisor d of n. Hence, $\sum \phi(d)$ accounts for all the elements in G.

82.

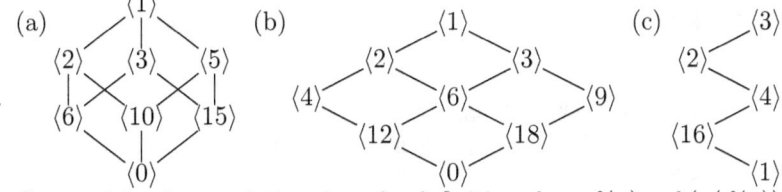

83. Composition is associative since, by definition, $h \circ g \circ f(n) = h(g(f(n)))$. The identity function is the identity permutation, and Theorem 3.9 guarantees that the inverse of a bijection is a bijection. Lastly, $n!$ is the number of ways one can order the elements in $\{1, 2, \ldots, n\}$.

84. It suffices to see that $(1, 2) \circ (1, 3) \neq (1, 3) \circ (1, 2)$ in S_n.

85. (a) 6 (b) 3 (c) 6 (d) 12

86.

$\langle(1,2,3)\rangle \quad \langle(4,5)\rangle$

87. D_4 and $(1,2) \circ D_4$ and $(1,4) \circ D_4$.

88. For D_3, let $R = (1,2,3)$, $F_1 = (2,3)$, $F_2 = (1,3)$, $F_3 = (1,2)$.

$\circ$	R	R^2	R^3	F_1	F_2	F_3
R	R^2	R^3	R	F_3	F_1	F_2
R^2	R^3	R	R^2	F_2	F_3	F_1
R^3	R	R^2	R^3	F_1	F_2	F_3
F_1	F_2	F_3	F_1	R^3	R	R^2
F_2	F_3	F_1	F_2	R^2	R^3	R
F_3	F_1	F_2	F_3	R	R^2	R^3

89. It is obvious for rotations. If F is a reflection, observe that if x and y are adjacent vertices, then $F(x)$ and $F(y)$ are still adjacent, but with reversed circular ordering.

90. A rotation circularly shifts the vertices of the n-gon, while a reflection shifts as well as reverses their ordering. So applying any number of rotations, or an even number of reflections, keeps the circular ordering—while an odd number of reflections reverses it.

91. Let $D_n = \{R, R^2, \ldots, R^n, F_1, F_2, \ldots, F_n\}$. Then $|F_m| = 2$ for each reflection, and $|R^m| = n/\gcd(m, n)$ for the rotations.

92. Being a reflection, hence self-inverse, $F \circ R = (F \circ R)^{-1} = R^{-1} \circ F$. If abelian, then $R = R^{n-1}$ by the cancellation law. But R and R^{n-1} are distinct group elements of $\langle R \rangle$, unless $n \leq 2$.

93. The rotations form a cyclic subgroup $\langle R \rangle$ of D_n. If H is a subgroup of odd order, then $H \subseteq \langle R \rangle$ since a reflection has order 2, not a divisor of $|H|$. As a subgroup of a cyclic group, H is cyclic.

Chapter 4

1. 28

2. $20 \times 3 \times 8 \times 999 = 479520$

3. Select the unit digit first, then left to right: $5 \times 8 \times 8 \times 7 \times 6 = 13440$.

4. $(3 \times 8 \times 7 \times 6) + (5 \times 9 \times 8 \times 7 \times 6) = 16128$

5. Write each composition using n-digit ones and zeros, e.g., $2+5+1+1 = 110000010$. Since we have 2 choices for each digit except the first, which must be 1, there are 2^{n-1} such numbers.

6. (a) 8 (b) 16 (c) 96 (d) 9

7. If d is the largest odd factor of n, then $n = 2^e \times d$ for some $e \geq 0$. In our range, $d \in \{1, 3, 5, \ldots, 1999\}$, so with 1001 integers, two will have

identical value of d, say $2^e \times d$ and $2^f \times d$. The larger of these two is a multiple of the smaller.

8. 29
9. (a) 160 (b) 57 (c) 943 (d) 948
10. (a)
11. (a) 215 (b) 78 (c) 922 (d) 948
12. $|A \cup B \cup C \cup D| = |A| + |B| + |C| + |D| - |A \cap B| - |A \cap C| - |A \cap D| - |B \cap C| - |B \cap D| - |C \cap D| + |A \cap B \cap C| + |A \cap B \cap D| + |A \cap C \cap D| + |B \cap C \cap D| - |A \cap B \cap C \cap D|$
13. $246 + 49 + 9 + 1 = 305$
14. (a) 120 (b) 6 (c) 834 (d) 39486
15. (b)
16. Permute the consonants first, then each vowel may be inserted in between, first, or last: $5! \times 6 \times 5 \times 4 = 14400$.
17. (c)
18. $P(15, 4) = 32760$
19. $P(23, 5)/5 = 807576$
20. From Amira's perspective, to her right is 4 choices, leaving 3 choices to her left. The remaining three people include Elias and are free of further restrictions. Thus, $4 \times 3 \times 3! = 72$.
21. (c)
22. Let the largest term be $n + k$. Then $C(n + k, k)$ is the product of the k consecutive integers divided by $k!$.
23. (a) 330 (b) 386 (c) 386 (d) 1013
24. They constitute the total number of subsets of a set with n elements.
25. For distinct pairs, there are $C(7, 2) = 21$ dominoes. With the seven doubles, the total is 28.
26. There are $C(10, 4) + C(10, 5) + C(10, 6) = 672$ sets of four to six exam sessions, so $\lceil 7469/672 \rceil = 12$ students must belong together in one set.
27. (a) 171 (b) 78 (c) 216 (d) 231
28. The first must be positive, thus $C(12, 9) = 220$. This quantity includes the 5-digit 10000, so the correct answer is 219.
29. (b)
30. By way of its complement, $231 - 65 = 166$.
31. (a) 1/18 (b) 1/6 (c) 11/36 (d) 7/12
32. 4/15
33. (b)
34. Out of 1024, (a) 252 (b) 240 (c) 1013 (d) 175
35. Fix one person; to the right will be $5 \times 4 \times 4 \times 3 \times 3 \times 2 \times 2 \times 1 \times 1$. With 9! in the sample space, the probability comes to 1 in 126.
36. Out of 256, (a) 56 (b) 32 (c) 247 (d) 65
37. $(1/2)^3 + (1/2)^3 - (1/2)^6 = 15/64$
38. $1 - (365/365)(364/365)(363/365) \cdots (343/365) = 0.507297$
39. Out of 6,497,400, (a) 358800 (b) 1092 (c) 6417075 (d) 186745

40. (d)

41. (a) 40 (b) 3744 (c) 54912 (d) 123552

42. Maybe later when we have time to play.

43. Both equal $\frac{n!}{(k-1)!\,(n-k)!}$. The summation yields $n\sum\binom{n-1}{k-1} = n2^{n-1}$.

44. $(x+y)^{n+1} = (x+y)^n(x+y) = (x+y)\sum\binom{n}{k}x^{n-k}y^k = \sum\binom{n}{k}x^{n+1-k}y^k + \sum\binom{n}{k}x^{n-k}y^{k+1}$. Combine like terms,
$x^{n+1} + (\sum(\binom{n}{k+1}) + \binom{n}{k})x^{n+1-k}y^k) + y^{n+1}$, then use Pascal's formula.

45. $x^8 + 8x^7y + 28x^6y^2 + 56x^5y^3 + 70x^4y^4 + 56x^3y^5 + 28x^2y^6 + 8xy^7 + y^8$

46. $2^{n+1} = 2^n + 2^n = \sum\binom{n}{k} + \sum\binom{n}{k} = 1 + \sum_{k=0}^{n-1}\binom{n}{k+1} + \sum_{k=0}^{n-1}\binom{n}{k} + 1 = 1 + \sum_{k=0}^{n-1}\binom{n+1}{k+1} + 1 = \binom{n+1}{0} + \sum_{k=0}^{n-1}\binom{n+1}{k+1} + \binom{n+1}{n+1} = \sum_{k=0}^{n+1}\binom{n+1}{k}$.

47. (a) Just let $x = 1$ and $y = -1$ in the binomial theorem.

(b) Gather the odd k, i.e., negative terms, across the equal sign.

(c) The equal halves sum to 2^n, so each half is 2^{n-1}.

(d) We select any subset from $\{x_1, x_2, \ldots x_n\}$. Each x_i has two possibilities: chosen or not chosen, hence 2^n subsets. But to determine an even or odd number, x_n will lose its options, making it 2^{n-1}.

48. We choose n from n men and n women. To have k men, there are $\binom{n}{k}\binom{n}{n-k} = \binom{n}{k}^2$ ways. Sum over $0 \le k \le n$ to get $\binom{2n}{n}$ ways in all.

49. It is true for $k = 0$. Assuming $\binom{p-1}{k-1} - (-1)^{k-1}$ is a multiple of p, by Theorem 4.14, $\binom{p-1}{k} + (-1)^{k-1} = \binom{p}{k} - \binom{p-1}{k-1} + (-1)^{k-1}$ is also a multiple of p. We are done since $(-1)^{k-1} = -(-1)^k$.

50. (a) 2^n (b) $3 + 2^n$ (c) $4n + 7$ (d) $5 + n^2$

51. (c)

52. $S_n = S_{n-1} + S_{n-2} + S_{n-3} \to 1, 2, 4, 7, 13, 24, 44, 81, 149, 274$.

53. $F_n = 2^{2^n} + 1 = (2^{2^{n-1}})^2 + 1 = (F_{n-1} - 1)^2 + 1 = F_{n-1}^2 - 2F_{n-1} + 2$

54. (a) $2(3^n) + (-2)^n$ (b) $2^{n+1} - 1$ (c) $3(2^{2n-1}) - 2^{n-1}$

(d) $(\frac{13+\sqrt{13}}{26})(\frac{1+\sqrt{13}}{2})^n + (\frac{13-\sqrt{13}}{26})(\frac{1-\sqrt{13}}{2})^n$

55. $S_n = S_{n-1} + 2S_{n-2} = \frac{1}{3}((-1)^n + 2^{n+1})$

56. $1 + 2^n - 2(-1)^n$

57. $(2 - n)3^n$

58. It is true for $n = 0$ and $n = 1$. If it holds up to $n - 1$, then $S_n = 4S_{n-1} - 4S_{n-2} = 4(2^{n-1} - 2^{n-2}) = 4(2^{n-2}) = 2^n$.

59. (a) $S_n = 8S_{n-1} - 24S_{n-2} + 32S_{n-3} - 16S_{n-4} = (c_3n^3 + c_2n^2 + c_1n + c_0)2^n$

(b) $S_n = 27S_{n-2} - 54S_{n-3} = (c_1n + c_0)3^n + d(-6)^n$

(c) $3S_{n-2} - 3S_{n-4} + S_{n-6} = c_2n^2 + c_1n + c_0 + (d_2n^2 + d_1n + d_0)(-1)^n$

(d) $4S_{n-1} + 5S_{n-2} - 36S_{n-3} + 36S_{n-4} = (c_1n + c_0)2^n + d3^n + e(-3)^n$

60. (a) $1/(1+x)$ (b) e^x (c) e^{-2x} (d) $x/(1-x^2)$

61. $(1-x)/(1-px)$

62. $x(1+x)/(1-x)^3$

63. Expanding the product, observe that the coefficient of x^n is the number of non-negative solutions to $1x_1 + 2x_2 + 3x_3 + \cdots = n$, which equals the number of partitions of n into $1, 2, 3, \ldots$

64. (a) $x^4/(1-x^2)^4$ (b) $x^{10}/(1-x)^4$ (c) $1/(1-x)^3$
    (d) $x^3(1-x^{10})/(1+x)^2(1-x)^4$
65. $1/(1-x)(1-x^5)(1-x^{10})(1-x^{25})$
66. $\frac{1}{2}(n^2+3n+2)$
67. $(2-n)3^n$
68. $1+2^n-2(-1)^n$
69. (a) e^{-x} (b) $e^{x/2}$ (c) xe^x (d) $\frac{1}{2}(e^x-e^{-x})$
70. (a) $\frac{1}{8}(5^n+3^{n+1}+3+(-1)^n)$ (b) $\frac{1}{4}(5^n-1)$
    (c) $5^n-(n+2)4^n+(n+1)3^n$ (d) $(1+n+n^2)3^n$
71. We have $f''(x)-f'(x)-f(x)=0$, from which we obtain the solution
    $f(x)=(e^{x(1+\sqrt{5})/2}-e^{x(1-\sqrt{5})/2})/\sqrt{5}$.
72. Simply note that $|\frac{1}{2}(1-\sqrt{5})|<0.62$.
73. Observe that $F_0 F_1 \cdots F_{n+1} = (F_{n+1}-2)F_{n+1} = F_{n+1}^2 - 2F_{n+1} = F_{n+2}-2$ by Exercise 4.53.
74. (a) The induction step: $f_0 + \cdots + f_{2n+3} = f_{2n+2} + f_{2n+3} = f_{2n+4}$.
    (b) We know $f_0 + \cdots + f_{2n} = f_{2n+2}-1$. Subtract the odd terms, according to (a), to get $f_{2n+2}-1-f_{2n} = f_{2n+1}-1$.
    (c) Subtract the odd terms from the even. If n is even, $f_{n+1}-1-f_n = f_{n-1}-1$. If odd, $f_n-1-f_{n+1} = -f_{n-1}-1$.
    (d) $f_0^2 + \cdots + f_{n+1}^2 = f_n f_{n+1} + f_{n+1}^2 = f_{n+1}(f_n+f_{n+1}) = f_{n+1}f_{n+2}$
75. The induction step: $f_{m+n+1} = f_{m+n}+f_{m+n-1} = (f_m f_{n+1}+f_{m-1}f_n)+(f_m f_n + f_{m-1}f_{n-1}) = f_m(f_{n+1})+f_{m-1}(f_{n+1})$.
76. The sequence $f_n \bmod 3$ is periodic, repeating every eight terms, i.e., $0,1,1,2,0,2,2,1$. The zero residue occurs every multiple of four.
77. (a) If n is divisible by d, then f_n is divisible by f_d.
    (b) Since $\gcd(n,n+1)=1$, then $\gcd(f_n,f_{n+1})=f_1=1$.
    (c) If f_n divides f_m, then $f_n = \gcd(f_m,f_n) = f_{\gcd(m,n)}$. Hence $n = \gcd(m,n)$, and n divides m.
    (d) Because $f_5 = 5$.
78. Because f_n is divisible by 10 if and only if f_n is divisible by $2 = f_3$ and $5 = f_5$, i.e., if and only if n is divisible by 3 and 5.
79. $\begin{bmatrix} f_{n+1} & f_n \\ f_n & f_{n-1} \end{bmatrix}\begin{bmatrix} 1 & 1 \\ 1 & 0 \end{bmatrix} = \begin{bmatrix} f_{n+1}+f_n & f_{n+1} \\ f_n+f_{n-1} & f_n \end{bmatrix} = \begin{bmatrix} f_{n+2} & f_{n+1} \\ f_{n+1} & f_n \end{bmatrix}$
80. (d)
81. All four. With 323, they make the first five Fibonacci pseudoprimes.

Chapter 5

1. The exact same proof for graphs will do.
2. (a) $n(n-1)$ (b) $2mn$ (c) $2n-2$ (d) $2n$
3. By Euler's theorem, there are $n(n-1)/2$ edges. On the other hand, $1+2+3+\cdots+(n-1) = n(n-1)/2$.

4. (a)
5. (b)
6. (c)
7. (a) $n-1$-regular (b) only $K_{n,n}$ is n-regular (c) irregular (d) 2-regular
8. (a) $n-1, n-1, \ldots, n-1$ (b) $m, m, \ldots, m, n, n, \ldots, n$ if $m \geq n$
   (c) $2, 2, \ldots, 2, 1, 1$ (d) $2, 2, \ldots, 2$
9. (a) Yes (b) No (c) Yes (d) Yes
10. If the degrees are all even, draw n vertices with only loops. If some
    degrees are odd, pair them up each by an edge, since their number is
    even. (Why?) The remaining, all even degrees are again loops.
11. (b)
12. (a) two (b) four (c) three (d) two
13. (d)
14. It is true if $|V| = 2$ and assumed true for $|V| = n$. Let G be connected
    with $n+1$ edges. If $|E| \geq n$, we are done. Assume $|E| \leq n-1$, so
    $\deg G \leq 2n-2$. Since $\deg G = \sum \deg(v)$, there must be a vertex with
    $\deg(v) = 1$. Removing v and its edge, the subgraph is connected with
    n vertices, hence $n-1$ edges (at least) and n edges in G.
15. (c)
16. (a) $n = 2$ (b) m or $n = 1$ (c) all (d) none
17. (a) P_4 (b) $P_2 \sqcup P_2$ (c) trivial graph (d) $K_5 \sqcup K_4$
18. Draw the edges ab, bc, cd, be, ce.
19. Since $G \cup \overline{G} = K_n$, to have equally many edges, half that in K_n is
    $n(n-1)/4$, which is integer only if $n \bmod 4 = 0$ or 1.
20. $[A]_{ij} = 1$ if and only if (a) $i \neq j$ (b) $(i \leq m, j > m)$ or $(i > m, j \leq m)$
    (c) $|i - j| = 1$ (d) $|i - j| = 1$ or $n - 1$
21. $[A^2]_{ii} = \sum_{k \geq 1} [A]_{ik}$ since A is symmetric with 0 and 1 entries. This
    counts how many instances $v_i v_k \in E$, thus $\deg(v_i)$.
22. Where the ones are in each column:
    (a) Choose any 2 from n rows; $n(n-1)/2$ columns.
    (b) Pick one from top m rows, one from bottom n; mn columns.
    (c) Any 2 adjacent rows; $n - 1$ columns.
    (d) Like (c), plus another column with first and last rows equal 1.
23. (a) $\begin{bmatrix} 0 & 0 & 1 \\ 0 & 0 & 1 \\ 1 & 1 & 0 \end{bmatrix}$ (b) $\begin{bmatrix} 0 & 1 & 1 \\ 1 & 0 & 1 \\ 1 & 1 & 0 \end{bmatrix}$ (c) $\begin{bmatrix} 0 & 1 & 0 & 0 \\ 1 & 0 & 1 & 1 \\ 0 & 1 & 0 & 1 \\ 0 & 1 & 1 & 0 \end{bmatrix}$ (d) $\begin{bmatrix} 0 & 1 & 0 & 1 & 1 \\ 1 & 0 & 1 & 1 & 0 \\ 0 & 1 & 0 & 1 & 1 \\ 1 & 1 & 1 & 0 & 1 \\ 1 & 0 & 1 & 1 & 0 \end{bmatrix}$

24. (b)
25. (b)
26. (a) $n = 2$ (b) m or $n = 1$ (c) all (d) none
27. $P_2, P_3, P_4, K_{1,3}, P_5, K_{1,4}, \{ab, bc, cd, ce\}, P_6, K_{1,5}, \{ab, bc, cd, de, df\},$
    $\{ab, bc, cd, de, cf\}, \{ab, bc, cd, ce, cf\}, \{ab, bc, cd, be, cf\}.$

28. Adding an edge, by Theorem 5.6, make $|E| = |V|$, and by the same theorem, the new graph is no longer acyclic.

29. A leaf exists only in K_2, $K_{1,n}$, and P_n—exactly when the graph is a tree, by Exercise 5.26.

30. The case $n = 2$ is given by P_2. Proceed by induction. Let the sequence $d_1, d_2, \ldots, d_{n+1} > 0$ sum to $2n$. In particular, $d_1 \geq 2$ and we have seen that $d_n = d_{n+1} = 1$. Remove d_{n+1}, reduce d_1 by 1, and we get a sequence of n positive numbers that sum to $2n - 2$. Find a tree of this sequence and reattach the $n+1$st vertex to where $d_1 - 1$ refers to. The result is a tree of the desired sequence.

31. (d)

32. (a) 1 (b) 16 (c) 12 (d) 5

33. $D - A$ consists of all -1, except the diagonal entries are $n - 1$. Any cofactor C looks similar, only one size smaller: $(n - 1) \times (n - 1)$. Now add to top row each of the $n - 2$ other rows, to get a whole row of ones. Then add this new top row to each of the rest to get an upper triangular matrix with diagonal $1, n, n \ldots, n$. This procedure does not change $\det C$, which is n^{n-2}.

34. 404

35. (a) 4 (b) 0 (c) 0 (d) 10

36. (c)

37. (a) 1 (b) 2 (c) $n - 1$ (d) $\lfloor n/2 \rfloor$

38. By the theorem, $d(G) = 1$, 2, or 3. If $d(G) = 1$, G is complete and not self-complementary as $\overline{G}$ is trivial.

39. (a) $D = A$ (b) Like A, but every 0 off diagonal replaced by 2.
    (c) $[D]_{ij} = |i - j|$ (d) $[D]_{ij} = \min\{|i - j|, n - 1 - |i - j|\}$.

40. (a) $\begin{bmatrix} 0 & 2 & 1 \\ 2 & 0 & 1 \\ 1 & 1 & 0 \end{bmatrix}$ (b) $\begin{bmatrix} 0 & 1 & 1 \\ 1 & 0 & 1 \\ 1 & 1 & 0 \end{bmatrix}$ (c) $\begin{bmatrix} 0 & 1 & 2 & 2 \\ 1 & 0 & 1 & 1 \\ 2 & 1 & 0 & 1 \\ 2 & 1 & 1 & 0 \end{bmatrix}$ (d) $\begin{bmatrix} 0 & 1 & 2 & 1 & 1 \\ 1 & 0 & 1 & 1 & 2 \\ 2 & 1 & 0 & 1 & 1 \\ 1 & 1 & 1 & 0 & 1 \\ 1 & 2 & 1 & 1 & 0 \end{bmatrix}$

41. (a)

42. 133

43. If we treat each repeated edge as if there is a vertex in the middle, of degree two, the theorem applies without modification.

44. (a) walk (b) none (c) circuit (d) walk

45. (d)

46. Open walk: (a) $n = 2$ (b) m odd, $n = 2$, vice versa (c) all (d) none
    Closed walk: (a) n odd (b) m even, n even (c) none (d) all

47. (c)

48. $160 + (19 + 7) + (19 + 15) = 220$

49. (c)

50. (a) $n \geq 3$ (b) $m = n$ (c) none (d) $n \geq 3$

51. See page 166 for labeling. Suppose H is a Hamilton cycle. By symmetry, let the two copies of C_5 be joined via $1a \in H$. Since H is 2-regular, again by symmetry, we assume $ac \in H$ but $ad \notin H$, hence both $4d, bd \in H$. Now $5e \in H$, otherwise all $15, 45, be, ce \in H$, creating an impossible subcycle of 8 vertices in H. Symmetry again, we assume $be \in H$ but $ce \notin H$, hence $3c \in H$. Now $45 \notin H$, or else we have another impossible subcycle of 5 vertices. It follows, both $15, 34 \in H$, ending with another subcycle of 9 vertices. Hence H does not exist.

52. Each has degree 6, exceeding half of total 10 vertices. Similarly, $\overline{C}_n$ is $n - 3$-regular, where $n - 3 \geq n/2$ for $n \geq 6$. The special case $\overline{C}_5$ is self-complementary and is itself a cycle.

53. Unique Hamilton cycle of weight 95.

54. Unique Hamilton cycle of weight 624.

55. $18 + 18 + 23 + 23 + 25 + 25 + 31 + 31 = 194$

56. (a)

57. (a) $n = 2$ (b) all (c) all (d) n even

58. If $V_G = X \sqcup Y$, we may let $|X| = x$ and $|Y| = n - x$. The number of edges is maximized when G is complete bipartite, with $x(n - x)$ edges. Taking derivative, this quantity is maximum when $n - 2x = 0$, hence $x = n/2$ and the claim follows.

59. It is not.

60. P_2, $P_2 \sqcup P_2$, C_4, $P_2 \sqcup P_2 \sqcup P_2$, C_6, and $K_{3,3}$.

61. If 2-regular, Theorem 5.16 allows an Euler circuit. Such a walk will pass each vertex exactly once, hence is a cycle. Conversely, every cycle is 2-regular. Lastly, a cycle is bipartite if and only if of even length.

62. (b)

63. (a) n (b) 2 (c) 2 (d) $2 + (n \bmod 2)$

64. K_n for all n, and C_n for n odd.

65. (a) 4 (b) 3 (c) 3 (d) 3

66. Spread out the n white vertices along the x-axis and put the two black ones on the y-axis, one positive and the other negative. It works.

67. Put an imaginary vertex in the middle of every repeated edge. Euler's formula applies to the resulting plane graph: $v + r = e + 2$. Each imaginary vertex adds one to v, one to e, and none to r, hence the formula holds without them as well.

68. $K_{3,3}$ has 9 edges and no triangles, yet $9 > 2n - 4$, where $n = 6$ vertices.

69. Each has degree 6, contradicting Theorem 5.26. Similarly, $\overline{C}_n$ is $n - 3$-regular, not planar if $n \geq 9$. The special case $\overline{C}_8$ has 20 edges, too many for 8 vertices to be planar.

70. (b)

71. Color each old vertex black and the new ones white. It works.

72. (c)

73. I think three colors suffice.

74. Argentina, Bolivia, Brazil, Paraguay.

Index

www.ingramcontent.com/pod-product-compliance
Lightning Source LLC
Chambersburg PA
CBHW071715170526
45165CB00005B/2018